# 2016 Passenger Car and 2015 Concept Car Yearbook

**Other SAE books of interest:**

**2013 Passenger Car Yearbook**
(Product Code: ET-001)

**2014 Passenger Car Yearbook**
(Product Code: ET-002)

**Concept Car Year in Review: 2013**
(Product Code: ET-003)

**2015 Passenger Car and 2014 Concept Car Yearbook**
(Product Code: ET-004)

# 2016 Passenger Car and 2015 Concept Car Yearbook

From the Editors and Contributors of
*Automotive Engineering* magazine

SAE INTERNATIONAL®

Warrendale, Pennsylvania
USA

400 Commonwealth Drive
Warrendale, PA 15096-0001 USA

E-mail: mailto:CustomerService@sae.org
Phone: +1.877.606.7323 (inside USA and Canada)
+1.724.776.4970 (outside USA)
Fax: +1.724.776.0790



**ISSN Print 2375-611X**
**ISSN Online 2375-6128**
**SAE Order Number ET-005**
**http://dx.doi.org/10.4271/et-005**



**ISBN-Print 978-0-7680-8274-6**
**ISBN-PDF 978-0-7680-8275-3**
**ISBN-epub 978-0-7680-8277-7**
**ISBN-prc 978-0-7680-8276-0**

**To purchase bulk quantities, please contact:**
SAE Customer Service
E-mail: mailto:CustomerService@sae.org
Phone: +1.877.606.7323 (inside USA and Canada)
+1.724.776.4970 (outside USA)
Fax: +1.724.776.0790

**Visit the SAE International Bookstore at**
**books.sae.org**

# Table of Contents

# Preface

Each year, carmakers release new models with increasingly advanced technology intended to attract consumer interest and satisfy increasingly stringent government regulations. Some of these technologies are firsts-of-their-kind or leading–edge, and start trends that more companies soon follow-with the innovations becoming adopted across the board. In this book, we explore the technologies from an OEM and supplier perspective of the new production vehicle models released for the 2016 model year and the concept cars that point toward the future.

This *2016 Passenger Car and 2015 Concept Car Yearbook* continues the series of books that provide snapshots of the direction of the automotive industry. Each year, the editors of *Automotive Engineering* publish many articles focused on the technology and engineering innovations of new cars. These articles have been collected into this volume as a yearbook. In this singular publication, readers have a complete overview of the key advances that took place over the course of one model year.

Each car manufacturer has its own chapter in which its new models are explored in depth. With each new annual edition of the *Yearbook* series, trends and innovations are presented that make the automobile industry fascinating to the engineer and customer.

# 2016 Passenger Car Yearbook

# Audi A3 Sportback e-tron swells VW's plug-in hybrid choices

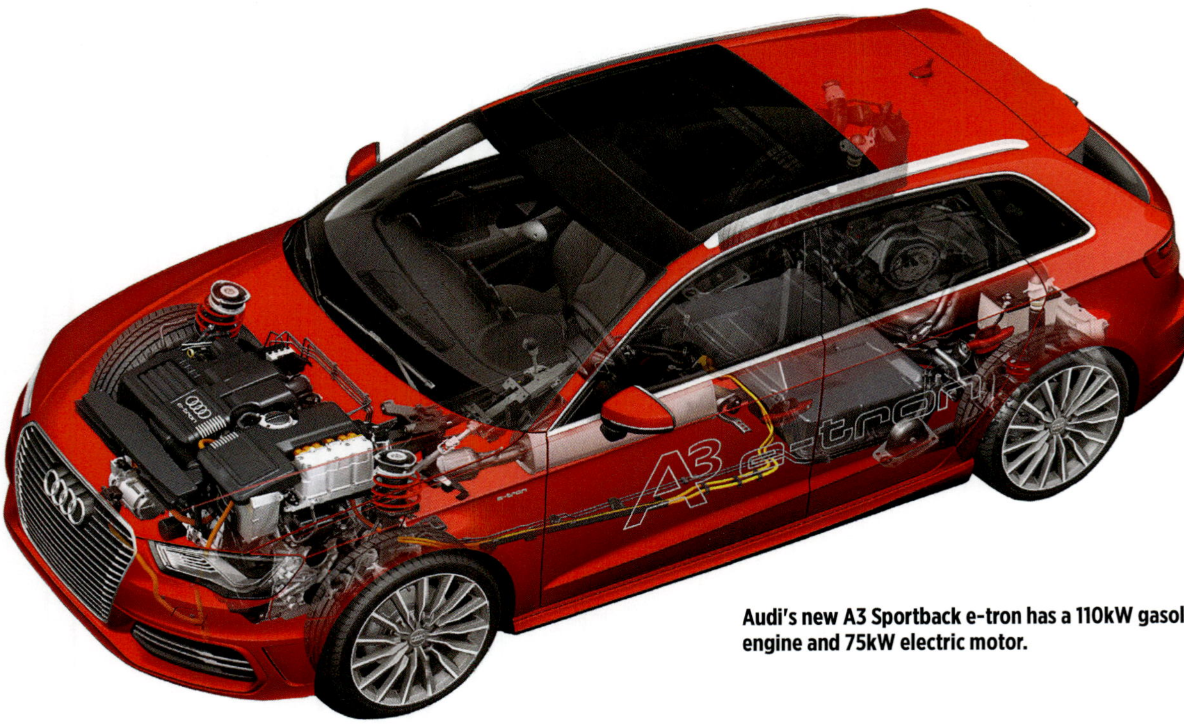

**Audi's new A3 Sportback e-tron has a 110kW gasoline engine and 75kW electric motor.**

"For Audi, the A3 Sportback e-tron plug-in hybrid marks another important step towards sustainable individual mobility," said Audi Head of Sales, Wayne Griffiths. But it is undeniably a small step, however worthy it may be. The car gets it's auto show debut in Detroit at the 2015 North American International Auto Show.

Persuading buyers that such technology meets their requirements at the right price remains a global challenge for the auto industry's sales and marketing expertise, and is a constant reminder that public mindsets are tough to change, no matter how sensible, sophisticated, and sustaining are the design and engineering programs supporting a solution.

Part of the persuasion campaign includes tax benefits and outright State-backed financial support. For example, in the UK, an A3 Sportback e-tron buyer gets a £5000 off-set from the government.

In keeping with the Volkswagen Group's policy of technology sharing, the Audi plug-in solution, now starting to be delivered to European customers, is very close to that in the Golf GTE (see http://articles.sae.org/13535/). The hybrid system comprises a 1.4-L TFSI 110-kW (148-hp) gasoline engine delivering 250 N·m (184 lb·ft) maximum torque, and a 75-kW electric motor having 330-N·m (243-lb·ft) torque capability.

Its official combined fuel consumption in accordance with the NEDC standard for plug-in hybrid vehicles is 1.5 L/100 km with $CO2$ emissions of 35 g/km. But testing in Europe by this *Automotive Engineering* editor over a wide mix of roads, saw an average of about 5.9 L/100 km, which is the reality of the concept. It can be argued, of course, that under selectable electric power alone, its range of 50 km (31 mi) makes it a ZEV for that distance. And some of that range can be "stored" for use in various journey phases through urban areas. The car's total range is 900 km (560 mi).

Purely from a driving aspect, the solution is very satisfactory in terms of performance (0-100 km/h in 7.6 s, top speed of 222 km/h) and, generally, regarding ride and handling, although the 125-kg (275-lb) weight of its 8.8-kW·h, eight-module, 96-cell lithium ion battery positioned under the floor beneath the rear bench seat makes itself felt when the car is driven quickly over winding roads. The battery sits in an aluminum housing.

Battery temperature is controlled via a liquid cooling system with its own low temperature circuit. Power electronics are positioned under the hood.

The A3 Sportback e-tron's disc-shaped electric motor is integrated between the flywheel and a bespoke designed six-speed e-S tronic (dual-clutch) transmission.

**Instrument details of the A3 Sportback e-tron.**

Vehicle weight is 1615 kg (3560 lb) unladen (EU) but with driver. The car comes standard with a relatively high specification. Externally, the A3 Sportback e-tron differs little from a regular Sportback.

The battery can be fully charged in some 2.5 h via a 3.6-kW charger. Energy recovery is achieved during braking. The vacuum brake servo has an additional supply connection served by an electric vacuum pump. Low to moderate braking forces can be achieved depending on the driving mode selected, via the electric motor as it operates as an alternator, putting expended energy back into the battery. The battery has an 8-year life guarantee.

All relevant components within the high-voltage network of the car have no conductive connections to other parts of the vehicle, claims Audi. In a crash situation sufficiently serious to initiate belt tensioning and airbag deployment, the high-voltage network is instantly disconnected from the power supply.

The A3 Sportback e-tron's four-ring grille badge slides to reveal the charger socket.

The conventional A3's rev counter is replaced by a "powermeter" to display overall system power and there is a battery status display.

Audi states that it is planning many online services for its electrified cars, providing various activating processes including charging processes. ■

*Stuart Birch*

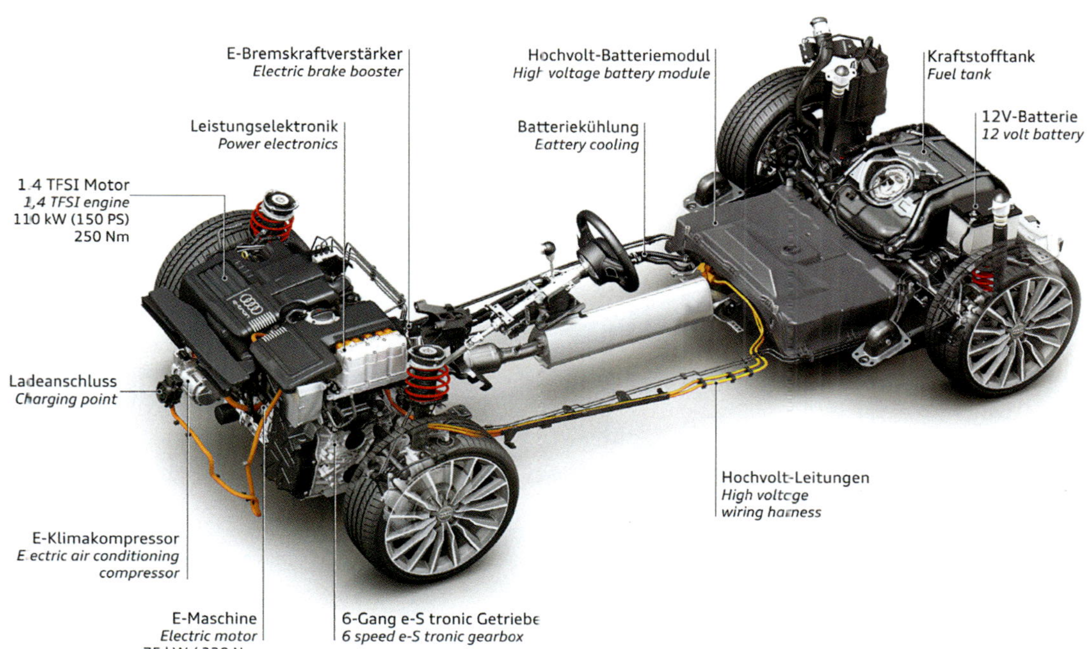

# Audi chooses high technology, cautious design evolution for new A4

**The fifth-generation Audi A4's road presence is about evolution and aero efficiency with a best Cd of 0.23. Compared to the previous generation, all panels have been changed and the grille redesigned.**

Audi's just-revealed 2016MY A4 is a classic example of automotive evolutionary design. Its aesthetics are not quite ripe for a game of "spot the difference" between the new fifth and previous fourth generation of this hugely successful model line, but they are close to it, even though every panel has been changed.

What this car is about, is evolution, technology, and systems' development; development of practically every aspect of its being (including new engine technology), without risking the sort of image step-change that Mercedes-Benz took in 2014 with the rival C-Class, a decision that has paid huge dividends in terms of sales and brand appreciation.

The new A4's reveal (sedan and Avant) in Germany ahead of its public unveiling at September's Frankfurt Motor Show was not accompanied by any official public comment by Audi's main Board members, who presumably are allowing the car to "speak" for itself.

In a sense it does so: check out the interior space and it has increased; try the headlights and (optional) matrix LEDs, developed in conjunction with Hella, will illuminate; interrogate the MMI and natural language voice control will respond; and the car gets Audi's virtual cockpit technology that it is rolling out to embrace its wider model range.

The A4's suspension has a redesigned five-link system front and rear to provide improved levels of ride comfort and sharper handling; one of its transmission options incorporates freewheeling; engine choice includes units providing exceptionally low fuel consumption and CO2 figures; and the A4 sedan is highly aerodynamically efficient, with a Cd of only 0.23.

Perhaps Audi reckons it has a classic design signature (such as the 911, Beetle, Mini) that can just be massaged every few years as technology is updated and efficiency improved. But it is what is inside the wrapper, rather than the wrapper itself, that really distinguishes the A4's advance.

While the A4, using the MLB (modular longitudinal) EVO platform, gets a plethora of advanced electronic infotainment and connected systems, the fundamentals of improving a vehicle remain vital, with weight reduction still a salient issue; the 120-kg (265-lb) improvement against the fourth-generation car is significant.

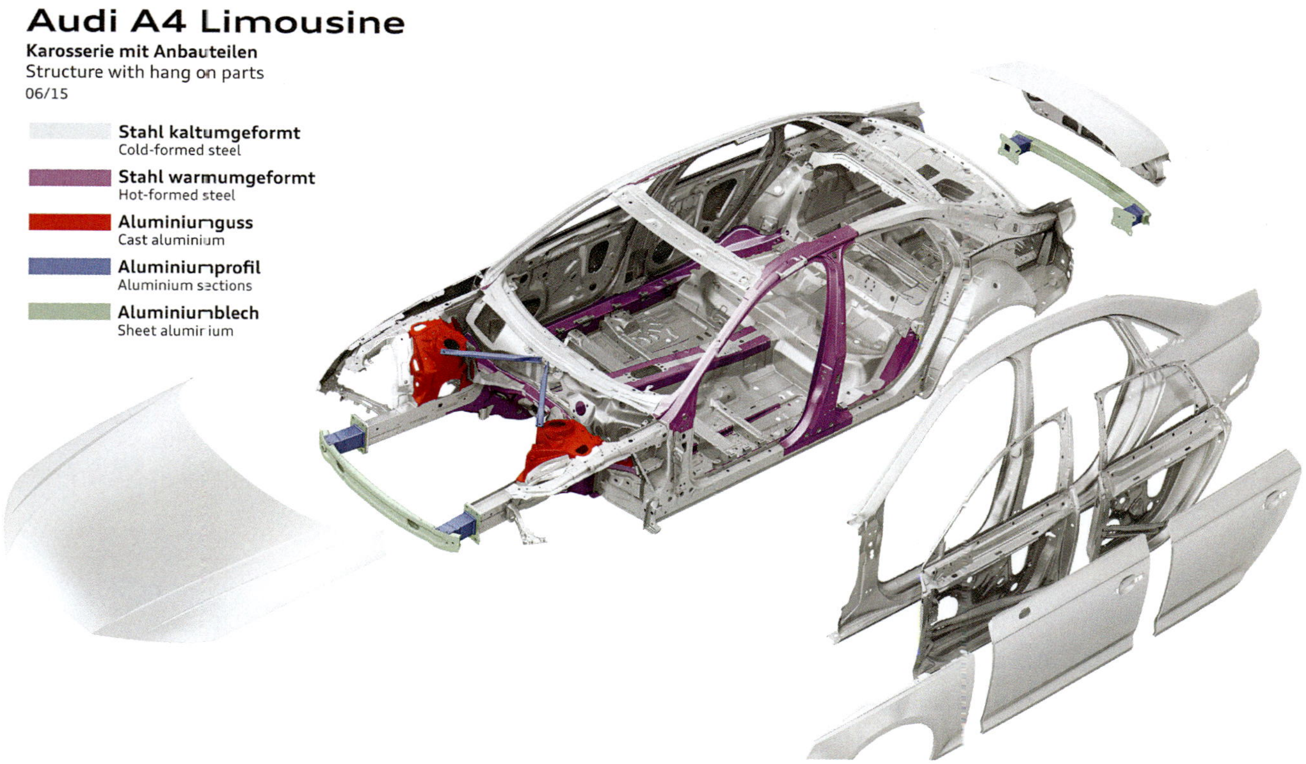


## Intelligent materials

The 1.4 TFSI gasoline base sedan has a 1320-kg (2910-lb) mass empty and without driver. The car's body is 15 kg (33 lb) lighter than its predecessor thanks to "geometrical lightweight construction and an intelligent combination of materials"—mainly aluminum and high-strength steels.

The module crossmember under the dashboard is now of extruded and sheet aluminum, the front crossmember an extruded profile. The mounts for the front MacPherson struts are highly integrated aluminum castings for an 8-kg (18-lb) savings. The construction is said to allow a very stiff connection between the upper ends of the struts and the car body for optimal driving dynamics.

Hot-stamped components form the high-strength, crash-proof backbone of the passenger compartment. They strengthen the transition from the front of the car to the interior, the frontal area of the roof frame, the B-pillars, the doorsills, and parts of the floor—and constitute 17% of the body structure. Interior weight watching includes lightweight seats.

The A4 sedan's length is 4726 mm, 25 mm (1 in) longer than the previous model. Width expands 16 mm (0.6 in) to 1842 mm (72.5 in). Height remains 1427 mm (56.2 in). Wheelbase is stretched 12 mm (0.5 in) to 2820 mm (111.0 in). The Avant is 1 mm (0.04 in) shorter and 7 mm (0.3 in) taller than the sedan but all other dimensions are the same.

Audi claims up to a 21% fuel consumption improvement for its latest engine range. Initially it spans from four-cylinder 1.4-L gasoline to six-cylinder 3.0-L diesels; power output spreads from 110 to 200 kW (148 to 268 hp).

The A4 2.0-L TDI (diesel) ultra (efficiency) sedan delivers a very frugal fuel consumption of 3.7 L/100 km, with CO2 emissions of 95 g/km. But gasoline technology developments are also bringing impressive results.

## Ultra solution

Audi's gasoline 2.0-L TFSI ultra model produces 140 kW (188 hp) and maximum torque of 320 N·m (236 lb·ft) from 1450 to 4200 rpm. Performance figures include 0-100 km/h (0-62 mph) in 7.3 s and a top speed of 240 km/h (149 mph) for the sedan with an S tronic transmission. NEDC (New European Driving Cycle) fuel consumption is a very good 4.8 L/100 km and CO2 emissions are 109 g/km.

Audi reveals that new technology is contributing to this efficiency. The 2.0 TFSI (see http://articles.sae.org/14140/) employs a new combustion method with shorter compression and longer expansion phases, and raised compression is designed especially for partial load, the intake valves closing much earlier than usual. Together with increased pressure in the intake manifold, this reduces throttling losses.

Because of the shortened compression phase, the compression ratio is upped from 9.6:1 to a heady 11.7:1. So in the compression phase, the engine has only to compress as much gas as a 1.4 TFSI. In the expansion phase, it gains from the high compression ratio, the resulting higher level of pressure during combustion further increasing the engine's efficiency.

**Design and engineering focus on aerodynamics has given the new A4 Avant a Cd figure of 0.26.**

For the 2.0 TFSI air/fuel mixture to swirl sufficiently despite the short intake time, the combustion chambers, piston recesses, intake ducts, and turbocharging are specially adapted. Under higher loads, the Audi valvelift system opens the intake valves later, resulting in a higher charge, which ensures good power and torque delivery. Injection pressure has been increased to 250 bar (3.6 ksi).

On the transmission front, the Multitronic CVT system has been dropped and choices now embrace a redeveloped six-speed manual, seven-speed dual-clutch S tronic, and eight-speed Tiptronic conventional automatic.

## Playing the Cd game

While weight reduction and advanced powertrains are at the epicenter of Audi's efficiency, aerodynamics have taken a very significant step with the new car, as Audi shows Mercedes-Benz what it, too, can do in that discipline. The sedan's 0.23 Cd is complemented by the Avant's 0.26.

Audi explains how some of it is accomplished by attention to fine detail.

The outer lines of the air inlets integrate additional openings, guiding some of the air stream over the wheel well, where it flows past wheels that have also been aerodynamically optimized.

The Avant gets a roof edge spoiler with a positive downward slant, and narrow trim on both sides of the back window aid airflow.

The underside of the engine compartment of both A4 versions is sealed from the road, and broad trim under the passenger cell and the luggage compartments protects the metal. The wishbones on the rear axle are enclosed and small spoilers guide the airflow at various areas of the car.

The A4 TDI ultra gets a controllable cool-air inlet. A frame installed behind the radiator grille houses two blinds that can be activated independently. They are closed as the car starts so that air resistance stays as low as possible; when air is needed to cool the engine, first the lower and then the upper blind opens.

The accent on aerodynamic efficiency has also been aimed at providing the A4 with a notably quiet cabin, particularly at high cruising speeds.

## Electronic upgrade

However, it is the level of cabin and electronic driver support technology that increasingly establishes a new car's status.

For the A4 it includes, depending on version and trim level, the Audi virtual cockpit with 12.3-in LCD screen with high resolution graphics; predictive efficiency assistant; traffic jam assist; rear cross-traffic assist; collision avoidance assist; and turn assist. There is an Audi smartphone interface, Audi Drive Select, and a second-generation modular infotainment platform with a 7-in color MMI (Multi Media Interface) monitor.

The infotainment system includes MMI Navigation Plus, MMI Touch, and Virtual Cockpit developed with Harman and Bosch. The modular infotainment system processor, called MIB II, is a Nvidia quadcore Tegra 30 chip.

An optional Audi tablet, available as a rear-seat entertainment device that's portable, employs a Google Android operating system and Nvidia Tegra 40 processor. ■

*Stuart Birch Q7*

# Audi lightens new generation Q7

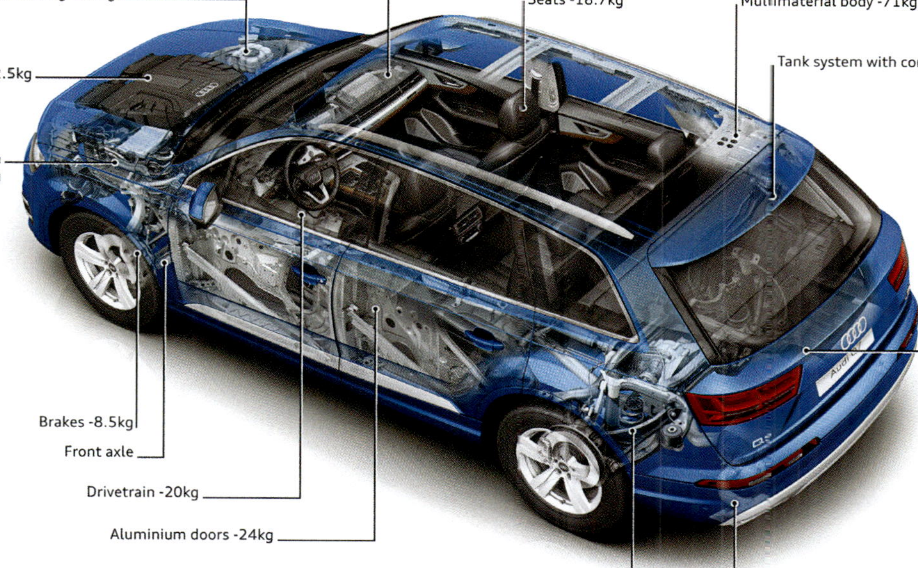

**Audi technical development boss,Prof. Dr. Ulrich Hackenberg, says the new technology-loaded Q7 is a statement of his company's competence.**

Audi's technical development boss, Prof. Dr. Ulrich Hackenberg, describes the new-generation Q7 SUV as representing a statement of the company's competence as he lists succinctly the salient points of its design assets: "Diversified drivetrain technologies, intelligent multimaterial lightweight construction—'We have achieved a weight reduction (compared to the outgoing Q7) of up to 325 kg (717 lb),' path-finding efficiency, intuitive connectivity, and comprehensive MMI and electronics features."

There is more to come, said Hackenberg: "In our strategy to electrify our entire product portfolio, we will introduce the highly efficient Q7 e-tron quattro later this year, with plug-in electric drive giving an electric range of more than 50 km (31 mi). It will be the first [plug-in hybrid] worldwide with permanent all-wheel drive powered by a six-cylinder 3.0-L diesel engine."

The e-tron Q7 will have a CO2 equivalent figure of 50 g/km.

Other significant Q7 technologies singled out by Hackenberg include matrix-LED headlights, advanced infotainment systems, "natural voice" recognition capability, Audi tablet with Google Android operating system, smartphone integration, and 3D sound system. It uses Audi's modular longitudinal platform (MLB), the matrix solution for cars with north-south mounted engines. The Q7 has all-wheel steering.

The Q7 technology onslaught is a sign of more to come, Audi Chairman Rupert Stadler pledged: "We will invest €24 billion—nearly $30 billion—over the next five years. The lion's share of that money will flow into the exciting project of series release for piloted driving; outstanding features like new standards in the connected car; an ambitious roadmap to drastically reduce CO2 emissions, and the further rollout of our new large-model initiative."

Audi's weight reduction embraces a raft of areas, from the electrical system to the luggage compartment floor.

**The Q7 has an impressive (for an SUV) Cd figure of 0.32.**

In 3.0-L 200-kW (268-hp) turbodiesel form, the Q7 now has a curb weight of 1995 kg (4398 lb), which contributes to a fuel consumption, compared to the previous generation Q7, improved by up to 23%. The 325 kg saved is the approximate equivalent of a grand piano, says Audi; an unusual comparison that seems somewhat out of tune with automotive engineering.

As usual for any weight cutting campaign, the entire Q7 design was analyzed. The body sees 71 kg (157 lb) saved via the use of aluminum and hot-shaped ultra-high-strength steel for the occupant cell. Aluminum castings, extruded sections, and panels are used for the car's front and rear end as well as its superstructure. Doors are aluminum, saving 24 kg (53 lb), as are the front fenders, hood, and rear hatch.

Three torsion rings are used to reinforce the Q7's front end, the area around the C-pillars, and the rear hatch cutout to improve static and dynamic body stiffness.

Also helping weight saving are slightly reduced dimensions compared to the previous model, with a length 15 mm (0.6 in) shorter at 5.05 m (16.6 ft), 15 mm less width at 1.99 m (6.5 ft), and a wheelbase shortened by 10 mm (0.4 in) to 2.99 m (9.8 ft). Despite these changes, the cabin is slightly longer and headroom is improved by up to 41 mm (1.6 in).

Improved aerodynamics complement reduced weight, with the Q7 achieving a very good SUV Cd of 0.32. The underbody is enclosed and at least one version of the Q7 gets active (via an energy controller) louvers between the front grille and the top of the water radiator.

Seating configuration includes an optional third row of two individual, certified Group 3 child seats. Interior comfort points include a newly developed two-zone (optional four-zone) air conditioning system using fewer buttons and controls. The front passenger area of the instrument panel incorporates a continuous air vent strip described by Audi as being "draft free."

Detailing the new car, Hackenberg stressed the significance of its latest generation MMI operating concept to control what is termed "the multitude of convenient and innovative functions" in the Q7.

Displays appear in the central MMI panel, which, when started, rises from the instrument panel into the driver's sight. It has a "full touch" surface, with each touch accompanied by an acoustic and haptic response. A large touchpad is also fitted.

A significant advance is a voice-control system that understands everyday sentence construction rather than definitive "orders," with hundreds of command variations possible for each function.

Like the Audi TT sports car, the Q7 is available with a virtual cockpit incorporating a 12.3-in instrument panel display (see http://articles.sae.org/13091/) that gives the driver the ability to switch from virtual analog instruments to navigation or custom information.

The system uses a modular information platform (designated MIB) in its second generation, which includes a quad-core T30 processor supplied by Audi partner Nvidia.

Sound systems include Bose with 3D sound and Bang & Olufsen Advanced Sound System, also with 3D, speakers in the A-pillars providing the added aural dimension of height. The Bang & Olufsen system drives 23 speakers including subwoofers, with 1920 W of amplifier power.

Rear-seat passenger entertainment includes Audi tablets with 10.1-in screens as rear seat monitors, each using a Nvidia T4 processor and meeting crash safety requirements. The Audi tablet is connected to the MMI Navigation-plus system via Wi-Fi, also providing radio and media access. MMI Navigation-plus features an Audi Connect module that connects to the Internet via the LTE (Long Term Evolution) standard.

Passengers can surf via the Wi-Fi hotspot with download speeds up to 100 Mbit/s and send and receive emails using "a variety of applications."

The Q7 comes with a comprehensive network of electronic safety systems including cross-traffic assist to warn of approaching vehicles when reversing slowly. The car offers an exit warning system to alert occupants to a motor vehicle or cyclist approaching from the rear.

Other safety/convenience aids include a "trailer assistant" that allows a driver to reverse a trailer accurately and easily using the

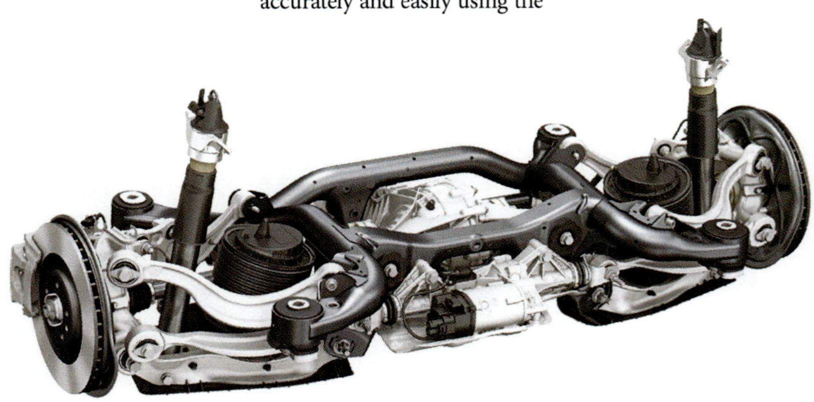

**A five-link rear axle, that incorporates just about everything including steering, is fitted to the top quattro version of the Q7.**

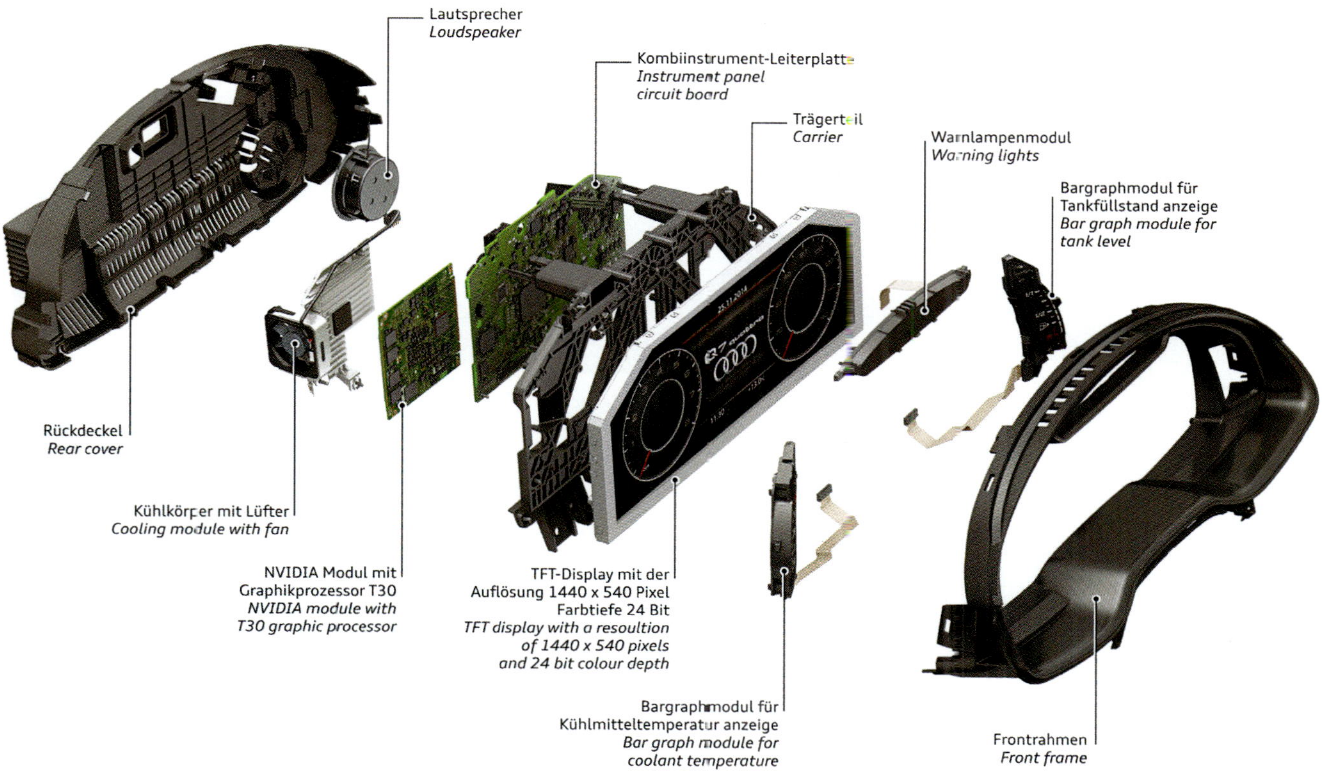


MMI rotary control to steer. The system also stabilizes a trailer being reversed in a straight line.

"Predictive efficiency assistant" uses the route data from the navigation system to alert the driver to approaching curves or urban speed limits that are beyond visual range. Linked to ACC (adaptive cruise control) and traffic-sign recognition, the system adapts the vehicle's speed to the route and the about-to-change situation, thus saving fuel to improve efficiency while enhancing safety. It can also apply a predictive element to operate freewheeling of the car's eight-speed transmission. Potential fuel saving could be as high as 10%.An aid to economy is what Audi describes as a "new vaporization system" that enables car propulsion with the engine at "extremely slow speeds"—less than 1000 rpm.

SAE Eye on Engineering Quick Take:

## Audi's inductive charging e-tron

Watch the video at: https://youtu.be/x2ILLTJP3xA

To charge an electric car, you typically plug it in to a power source. An alternative to plugging in is called inductive charging. In this episode of SAE Eye on Engineering, Senior Editor Lindsey Brooke looks at Audi's new Q7 e-tron plug-in hybrid with wireless charging technology. SAE Eye on Engineering can be viewed at https://youtu.be/x2ILLTJP3xA. It also airs in audio-only form Monday mornings on WJR 760 AM Detroit's Paul W. Smith Show.

Access archived episodes of SAE Eye on Engineering at www.sae.org/magazines/podcasts.

"Turning assistant" monitors opposing traffic at junctions when turning left, (or right depending on country) at low speeds, and can apply the brakes when necessary

All this electronic kit tends to overshadow the Q7's mechanical fundamentals, which include an initial choice of a 3.0-L TFSI (gasoline) unit with 245 kW (329 hp) in addition to the turbodiesel. The e-tron version will be on sale later this year.

Q7 suspension details include a five-link system front and rear instead of double wishbones, and there is a choice of air (including a sport version) or steel springs.

The all-wheel steering system incorporates an electric spindle drive; at low speeds, the rear wheels steer in the opposite direction to the front to aid parking and reduce the turning circle by about a meter; at higher speeds, the rears move in the same direction as the fronts. ■

*Stuart Birch*

# Audi unveils new R8 and all electric e-tron version

**The new Audi R8 is lighter and more powerful, while fuel consumption has been reduced.**

Although the second-generation Audi R8, which made its debut at the 2015 Geneva Motor Show, is similar in appearance to the outgoing model, new construction methods have reduced the weight of the car. Re-developed from the ground up, according to Audi, power will come from two variants of the 5.2-L FSI direct injection Audi V10 gasoline engine. Two further variants will also be avail able: an e-tron battery electric and an LMS race variant.

Power and torque have risen for the two V10-powered models. For the base V10 model, power output is a claimed 397 kW (532 bhp) with peak torque of 540 N·m (398 lb·ft) at 6500 rpm. This engine enables the car to accelerate from rest to 100 km/h (62 mph) in 3.5 s and on to a top speed of 323 km/h (201 mph). The more powerful V10 plus produces 449 kW (602 bhp) with peak torque of 560 N·m (413 lb·ft) at 6500 rpm. With this engine, the car will reach 100 km/h (62 mph) from rest in a claimed 3.2 s and a top speed of 330 km/h (205 mph).

The engine features indirect and direct fuel injection, for both lower fuel consumption and enhanced engine output depending on the situation. Fuel consumption is also benefited from cylinder on demand technology that deactivates injection and ignition to one of the banks of five cylinders and by engine stop-start. The dry-sump engine is assembled by hand at Audi's Hungarian engine plant.

Both V10 variants are equipped with a seven-speed S-tronic automated transmission, which offers three drive modes and a manual override. Four-wheel drive is fitted as standard. A launch control system manages full-throttle acceleration from rest. When power is not needed, the R8 is equipped with a coasting system. This is activated when the driver lifts off at speeds exceeding 55 km/h (34 mph). Both transmission clutches are opened and the car coasts to reduce fuel consumption.

Audi quotes fuel consumption of 11.8 L/100 km (19.9 mpg US) and CO2 emissions of 275 g/km for the R8 V10. This increases to 12.4 L/100 km (19.0 mpg US) with CO2 emissions of 289 g/km for the V10 plus. Audi claims a reduction in fuel consumption of up to 10% over the previous model.

For the revised four-wheel-drive system, the front axle is now equipped with an electrohydraulically activated multi-plate clutch in place of a viscous-coupling. Audi claims that drive torque can be directed wherever it is needed, with up to 100% torque distribution to either front or rear wheels. A mechanical differential lock is used for the rear axle.

Suspension is by aluminum double wishbones for all four wheels, and variable damping from magnetic ride shock absorbers is available as an option.

Electromechanical power steering is standard. Dynamic steering is offered as an option with variable ratio steering, which senses road speed and the settings chosen for the dynamic handling system.

Standard fitment on the base model are 19-inch wheels with 245/35 section tires in front and 295/35 section tires at the rear; 20-inch wheels are standard equipment for the R8 V10 plus and optional for the

**Performance mode is standard equipment on R8 V10 plus models, activated by a button on the steering wheel.**

base models. With 20-inch wheels, tire sizes are 245/30 at the front and 305/30 at the rear.

Braking equipment also varies according to model. The R8 V10 is fitted with steel brake discs, while the R8 V10 plus is equipped with carbon fiber ceramic discs.

Audi's drive select dynamic handling system gives drivers a choice of four modes: comfort, auto, dynamic, and individual. System behavior is governed by sensors monitoring engine characteristics, steering, transmission, and options such as the magnetic ride system, exhaust flaps, and dynamic steering—when fitted.

V10 plus models come with a new performance mode, available as an option for V10 models, activated from a button on the steering wheel. Drivers can choose from dry, wet, or snow settings, which will automatically engage the ESC electronic stability control system.

Audi engineers have devised a new method for spaceframe construction that reduces the weight of the body shell to 200 kg (440 lb), a claimed 15% lighter than before, for a total vehicle unladen weight of 1555 kg (3428 lb). Carbon fiber reinforced plastics are used in the construction of the B-pillars, central tunnel, and the rear wall. A framework formed from the front section, roof arch, and rear section is produced with cast aluminum nodes and aluminum

**A new R8 aluminum and composite spaceframe construction is lighter with greater torsional rigidity**

**Production R8 e-tron has greater battery capacity, while range has been almost doubled vs earlier developmental models.**

profiles, some using new alloys. Audi claims that torsional rigidity is improved by some 40%.

The R8 is 4.42 m (14.5 ft) long, 1.24 m (4.1 ft) tall, and 1.94 m (6.4 ft) wide or about 40 mm (1.6 in) wider than before. The wheelbase is 2.65 m (8.7 ft) long.

LED headlamps are standard equipment, with laser high beams offered as an option, which doubles the range of the high beam to 600 m (1970 ft). Where fitted, the laser high beam is made of four high-powered laser diodes that generate a blue laser beam with a wavelength of 45 nm. The blue light is converted to white using a phosphor converter with a color temperature of 5500 K, a color suited to human vision. The laser lighting is activated at speeds above 60 km/h (37 mph). A camera-based sensor system detects other road users and adjusts the light pattern to avoid dazzle.

The e-tron version benefits from the changes to the construction of the car. A T-shaped battery pack is integrated into the center tunnel and behind the cabin. Audi produces the lithium-ion cell, which offers a capacity of 92 kW·h compared with 49 kW·h for the previous developmental model. This provides a claimed electric range of 450 km (280 mi) instead of 215 km (134 mi).

Audi claims that energy density has been increased from 84 to 154 W·h/kg for the production model. The car can be charged using both direct and alternating current. Audi claims a full charge is possible in less than two hours.

The R8 e-tron offers claimed acceleration from rest to 100 km/h (62mph) in 3.9 s with top speed limited to either 210 km/h (130 mph) or 250 km/h (155 mph). ■

*Stuart Birch*

# Audi's tunnel of light leads to the future

**Audi's new drivable light tunnel is being used to research ever wider use of lighting systems for future models.**

Audi is seeing light at the end of its new research tunnel; and it is a laser. But not just any laser: this is a matrix laser that the Volkswagen brand is confident illuminates the way forward in lighting technology via an intensive light pattern for "virtually every situation" experienced on the road.

The company's engineers are focusing far ahead with R&D programs that include future cars saving energy by sharing road illumination. According to Audi's Dr. Wolfgang Huhn, Head of Development, Light/Visibility: "We are thinking in many directions, including machine learning, where self-learning algorithms continuously improve the illumination."

Light from matrix laser and matrix LED technologies will be linked to camera systems to control light with great precision, added Stephan Berlitz, Head of Audi's Lighting Innovations: "And where our field of vision ends, we can use Car-to-X technologies—i.e., information from other cars and infrastructure."

Audi pioneered the MatrixBeam technology—together with Hella and Automotive Lighting on the TT and ZKW (Q7)—over the past couple of years, and its new "tunnel" will be central to extrapolating that work and pushing the light envelope still further.

The tunnel is at the heart of company's new €4.5 million Lighting Assistance Center (LAC) at its Ingolstadt, Germany, headquarters. With a length of 120 m (390 ft), it is described as the largest drivable light tunnel in Europe. It has a street-like floor, a turntable and scale for vehicles, and a laser laboratory among its facilities. For most of its length it is 12 m (39 ft) wide and 5 m (16 ft) tall, but these figures expand over the final 25 m (82 ft) length to 18 and 9 m (59 and 30 ft), respectively.

The work in the tunnel and associated laboratories (in the basement of Audi's Center for Electrics and Electronics' building) embraces every aspect of car lighting, including not only safety but also aesthetics—something that Audi

Headlights, taillights, interior, and even body lighting are all on Audi's R&D lighting list.

regards as a very significant aspect of marque identity. Berlitz wants to take the emerging light technologies and arrange them into what he terms "a new aesthetic language." Various models are being given individually distinctive daylight running lights sufficiently subtle to indicate the type of vehicle.

Audi's motorsport program—particularly Le Mans—continues to play a significant role in lighting development for road and track. The racing R18 e-tron quattro and special edition road-going version of the outgoing generation R8 LMX were shown at Le Mans with a laser spot system that incorporates laser diodes linked to phosphor converters capable of creating white light. Earlier this year, Audi revealed matrix laser technology. This involves a laser spatially separated from the headlight. The system includes the use of over 400,000 separately controlled micromirrors breaking up the light into minute pixels. Each micromirror can be tilted by some 5000 times/sec. The system allows the creation of an almost infinite number of light patterns for almost every driving situation. It can also provide light markers on a road surface to help the negotiation of narrow lanes through construction work.

In the R8 LMX, the laser spot supplements the LED high beams.

Laser-spot capability is to be introduced on several Audi models. The system includes four powerful laser diodes in each headlamp module, each with a diameter of 0.3 mm (0.01 in), to generate a monochromatic and coherent blue laser beam having a 450-nm (18-µin) wavelength. A phosphor luminophore is then used to convert it into white light with a color temperature of 5500 K suitable for road vehicles. With its use of laser technology, Audi stresses that there is no risk to human eyesight.

The laser spot is dipped when an associated camera detects oncoming vehicles. The matrix laser headlight DMD (Digital Mirror Device) technology, commonly used in video projectors, is now in its pre-development phase at Ingolstadt for wider use in volume production vehicles.

Because lighting is used by Audi as a brand signature, a close association between designers and engineers is vital. Technical Development boss Prof. Dr. Ulrich Hackenberg has described lighting as "bringing technology to life and design differentiation." Berlitz added: "Our colleagues in Design help themselves to our ideas pool and in turn contribute their own suggestions."

The lighting systems of the future will also incorporate another differentiating feature: software, developed in-house by Audi.

**Dr. Wolfgang Huhn, Head of Development Light/Visibility at Audi, said: "With the lighting systems of the future, there will be a second differentiating factor besides design: the software."**

In the shorter term, R&D work may lead to production vehicles equipped with OLED technology for taillights. Organic light-emitting diodes are based on an organic paste-like material; when a voltage is applied, the molecules enclosed in the material release photons, causing the surface to illuminate.

At the light tunnel, Audi demonstrated OLEDs providing a 3D effect achieved by positioning OLED surfaces upright and one behind the other. "We are working in predevelopment with solutions where we apply OLED coatings directly to the vehicle body's sheet metal," said Berlitz. However, the new technology has its limitations. New materials are needed to allow OLEDs to be attached to a curved surface, and the material can only cope with temperatures of up to 80°C (176°F), so requiring effective thermal management. Audi has already succeeded in fusing light and the body in concepts and regards the results as another identity indicator.

For interiors, Audi is also investigating the use of small spots of light to enhance the surface effect of upholstery and decorative inlays. Light can provide what the company terms a "velvety soft" appearance to leather or make a CFRP (carbon fiber reinforced polymer) facing "reflectively glossy."

The eventual introduction of autonomous driving is expected to provide further roles for light technology. Audi demonstrated a steering wheel rim that uses green lighting that turns to red indicating to the driver when manual vehicle control is needed. An alert could also involve the lower windshield trim erupting into a red flashing line.

**"In the next 10 to 15 years, we will experience lighting innovations that we can barely even imagine today," says Audi's Head of Lighting Innovations, Stephan Berlitz.**

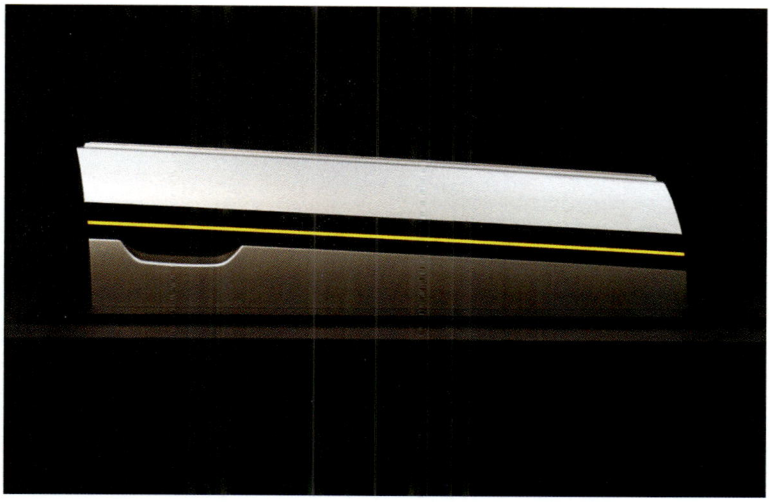

**Possible body flank lighting is another potential area of interest for Audi's lighting R&D teams.**

**The flexibility of OLED and other lighting technologies is significant in terms of vehicle identification as well as safety and end-user convenience.**

In addition to OLEDs, series applications of which could be less than 5 years away, further materials R&D work includes the use of MID (molded interconnected devices).

The technology is based on an innovative polymer containing an organic metal complex. MID lighting assemblies can be developed in any desired form, says Audi, adding that it represents a huge improvement compared to present 2D printed circuit boards.

The use of flexible silicone and foamed polymers also offer light treatments, the former allowing the creation of very small lenses with tight radii to provide refracting angles similar to glass lenses. Clasps for daytime running lights could be created using foamed polymers.

Optic fibers of polymer or quartz glass emitting tip, or whole length, light that can be woven into textiles are also on Audi's comprehensive light list.

Advanced lighting systems will require legal requirements to be modified in many markets, a factor which could slow introduction unless addressed soon. ■

*Stuart Birch*

# BMW lightens 7 Series, but bulks up on tech

**The 2016 BMW 7 Series' slippery new exterior suffers 15% less aerodynamic drag than its predecessor.**

Prestige car manufacturers are trading body blows among their top-segment models, with each ladling on more lavish content and more cutting-edge technology with each new introduction.

So it was for BMW at the 2015 Frankfurt Motor Show, where it pulled the cover off its 7 Series and released it to take on the Mercedes-Benz S-Class and Audi A8. We may not have yet reached the day when carbon-fiber chassis seen in sports cars have proliferated among these sedans, but the 2016 7 Series offers a step in that direction with a chassis featuring significant carbon-fiber reinforcement to a mostly steel structure. BMW terms this hybrid materials approach "carbon core."

The car's doors and trunk lid are aluminum, and use of that metal in the suspension, brakes, and wheels slashes unsprung weight by 15%.

"Its innovative combination of carbon-fiber structures with steel, aluminum, magnesium, and plastic means that the 750i xDrive is 130 kg (287 lb) lighter than the previous model," noted Klaus Frolich, BMW Board Member for Research and Development. "Lightweight design measures adding up to roughly 200 kg (440 lb) have enabled us to integrate additional comfort and safety features," he said. "This ensures that the 7 Series is the lightest luxury sedan in its segment."

Outside, the body shell is smoothed and airflowed aided by active measures such as grille shutters, an "air curtain" that steers airflow past the side of the wheel arch, while "air breathers" direct air into the front wheel arches through vertical intakes in the front fenders. These measures, along with a sealed underbody, contribute to a 15% reduction in aerodynamic drag.

Front and rear self-leveling air suspension lets drivers manually raise the car by 20 mm (0.8 in) when approaching steep entrances and automatically lowers is by 10 mm (0.4 in) when driving at highway speeds for reduced drag.

An active anti-roll bar is a feature increasingly seen on SUVs because it permits disconnection for wheel articulation when driving off road, then returns to its normal function of reducing body lean in turns on road. The new 7 Series uses one so it can disconnect the bar when the car is driving in a straight line to let each corner's suspension respond to bumps individually, without affecting the others. When turning, the anti-roll bar returns to its normal function.

The 7 Series also has a variable ratio steering rack and active rear steering, which combine to provide improved maneuverability.

What else? How about remote control self-parking? "This means the vehicle can maneuver itself in and out of a garage or parking spot without the driver behind the wheel," Frolich explained. The driver will use BMW's new Display Key, which appears to combine the appearance of a small smartphone with functions such as starting, the car's lock status, and even service notifications. Remote-control parking will not be available when cars reach dealers this fall, but is scheduled to follow soon afterward, he added.

Frolich boldly invited customers to try the 7 Series' new gesture-control system, an interface for the car's infotainment systems that reads the driver's gesticulating in its direction and interprets that to control functions like radio and phone. Considering

**The high-resolution infotainment display in the center of the 7 Series' dashboard is now a touch-screen for easier access to some functions.**

the mixed record of voice command systems and BMW's iDrive interface, it is worrisome to consider what gestures the computer system might induce.

BMW continues to refine its signature iDrive infotainment interface, finally combining touch screen inputs with its control knob despite a decade of insisting that fingerprints make touch screens undesirable.

A Bowers & Wilkins Diamond surround sound system features a 10-channel, 1400-W amplifier and 16 speakers to immerse occupants in sound. Meanwhile a new headup display projects information for the driver with an image on the windshield that is 75% larger than the previous system, with improved image quality and detail.

**The 4.4-L V8 nests its pair of Honeywell turbochargers inside the valley of the engine.**

BMW has announced three U.S. gasoline drivetrains for the 7 Series; a 445-hp (332-kW) twin-turbo 4.4-L V8 for the 750i, a 320-hp (239-kW) 3.0-L twin-turbo inline six for the 740i, and a 258-hp (192-kW) 2.0-L twin-turbo inline four and 83 kW electric motor plug-in hybrid-electric combination in the 740e. The hybrid's total system output is 243 kW. All three cars use eight-speed automatic transmission.

The V8 is a hot-vee design, nesting a pair of Honeywell twin scroll turbos in the valley of its vee. Compression for this iteration of the engine climbs a half-point to 10.5:1, with cooling optimized with separate flow to the heads and cylinder liners determined by a map-controlled water pump. System flow can be restricted to 10% of normal for faster warmup on cold starts.

BMW is debuting a new inline-six for 2016 featuring all-aluminum construction, variable valve timing, and direct gasoline fuel injection for a good combination of smoothness and power.

The plug-in hybrid 740e has a lithium-ion battery pack that can propel the car with no fuel consumption at speeds as high as 75 mph (120 km/h) or for distances as far as 23 mi (37 km). Charging is via standard household current, public charging station or BMW's i Wallbox. BMW claims maximum efficiency of 112 mpg for the plug-in hybrid.

Global markets will also enjoy the 730d, a 265-hp (198-kW) 3.0-L, 52-mpg fuel-sipping 3.0-L inline six-cylinder diesel. ■

*Stuart Birch*

# Cadillac ATS-V bulks up for track duty

**The ATS-V possesses the on-track poise expected of contenders in the premium sport sedan category.**

In developing the Cadillac ATS-V high-performance variant, General Motors' engineers were faced with an unfamiliar problem; a base car that was a little too light.

Not exactly, but that was the gist of the issue. After years of developing cars in which mass reduction was a seeming afterthought, GM has with recently developed models such as the ATS, on its Alpha platform, so thoroughly optimized on mass for the base model that the car was left needing a stronger foundation when more power and traction were added for the ATS-V model.

In both its two-door coupe and four-door sedan variants, the ATS-V received a raft of needed chassis reinforcements that produce a 25% increase in torsional rigidity, reported Cadillac Chief Engineer Dave Leone. This improvement was achieved through the installation of a front shock tower-to-plenum brace along with underhood V-braces, reinforcement to the rocker bulkhead, stamped steel bolt-on braces between the rear suspension cradle mount and the floor of the unibody, and (the centerpiece of the upgrades) a large aluminum platform beneath the front suspension subframe that the company terms a "shear plate" that connects the front suspension and its cradle more securely to the car's unibody.

The effect of that huge aluminum reinforcement plate is felt primarily on corner turn-in, where the ATS-V responds crisply rather than by twisting the car's structure, according to Leone.

With the platform suitably braced, GM's engineers beefed up the suspension components too. That includes replacement of bushings in the multi-link double-pivot MacPherson strut suspension with zero-compliance ball joints, stiffer front springs, and thicker front anti-roll bar that combine to produce a 50% increase in roll stiffness. The ZF Servotronic II steering rack is also stiffer than that on the regular ATS, which also contributes to improved steering feel.

At the rear, there are stiffer bushings, springs, and anti-roll bar along with stronger cradle mounts. Revised rear suspension geometry incorporates more anti-squat and has less migration of the rear roll center.

A standard electronic limited-slip differential helps put the power down on corner exits, while different-diameter halfshafts suppress axle hop under acceleration by not having the same frequency response to torque. The left side shaft, which is provided by GKN as an assembly with its constant velocity joints, is 55 mm (2.2 in) compared to 30 mm (1.2 in) for the standard-issue right shaft, for a 2.6:1 difference in rigidity between the two.

Brakes are the same Brembo six-piston one-piece front calipers with 14.5-in (368-mm) rotors at the front and four-piston rear calipers with 13.3-in (338-mm) rotors at the rear as were used on the heavier outgoing-generation CTS-V, which provided impressive stopping power without fade or increased pedal travel during track testing at the Circuit of the Americas in Austin, TX.

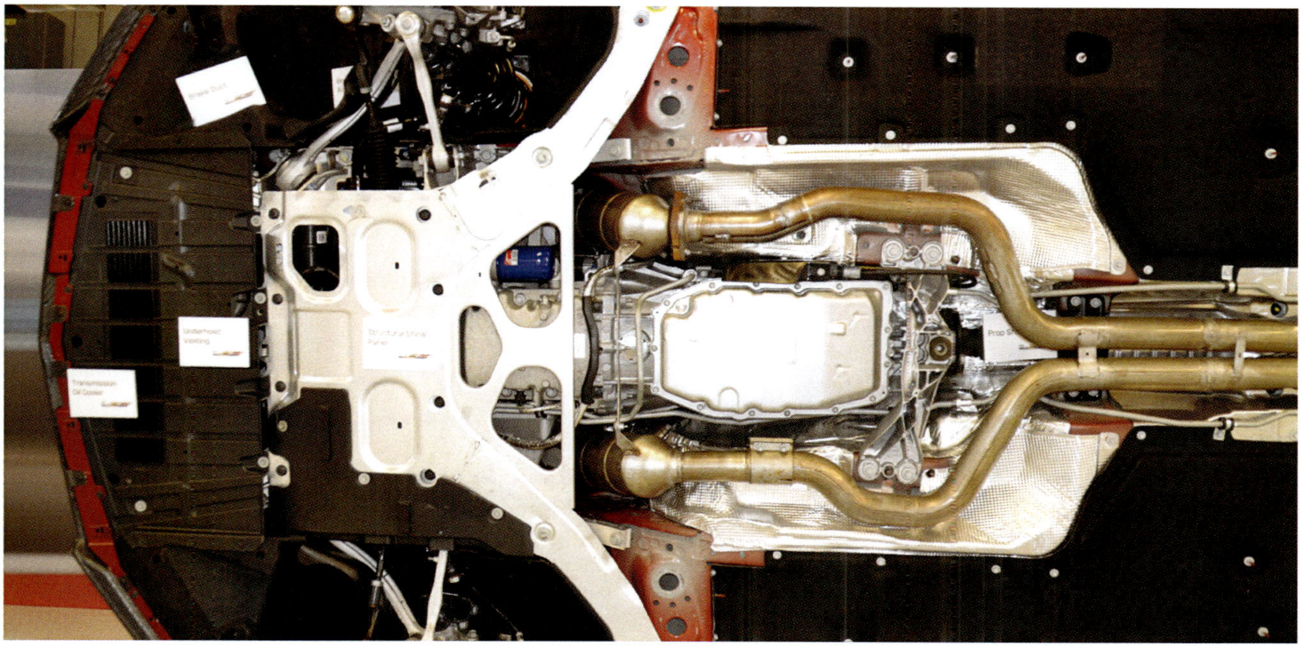

**The huge aluminum plate tying the front subframe to the unibody more securely provides the ATS-V with crisper response on corner turn-in, as the chassis doesn't absorb the initial forces by flexing instead of turning.**

Damping is by the BWI Group's Magnetic Ride Control that Cadillac brought to market. This latest iteration of the technology provides improved differentiation between its two modes, according to John Barrick, Program Engineering Manager for the ATS-V. "Now we have two true modes," he said. "Before it was more like one and a half modes." The increase in maximum damping is 40% greater than before.

The new system's quicker response time means that the damper adjusts itself for every inch the ATS-V travels at 60 mph (97 km/h), while the previous system covered three inches between adjustments to settings, Barrick added.

The 18-inch forged aluminum wheels help reduce mass and mount Michelin Pilot Super Sport tires molded with three distinctly different zones of compound. The outer shoulder is hardened to withstand the cornering force of hammering around tracks without burning the rubber off the outside of the tires. The middle section is optimized for wet traction, and the inner portion is sticky race-ready rubber.

On the track, the tires contributed to an impressive 1.25 g lateral acceleration in the turns.

Just as engineers attended to every detail under the skin, so did the designers when working on the skin itself. "All of the design elements have a purpose," said Andrew Smith, Executive Director of Cadillac Global Design. "They contribute to lift reduction, enhanced cooling, reduced mass, or all of the above."

Most obvious is the carbon-fiber hood, with its bare carbon heat extractor vent. The hood saves weight compared to a steel hood, and the extractor not only helps hold down engine bay temperatures, but the air exiting it atop the hood reduces lift at speed.

The enlarged openings in the front fascia feed more air to the various heat exchangers behind it, and the splitter on the bottom scrapes directs air from the road surface, minimizing lift. The ATS-V's fenders are widened to cover the wider wheels and tires, and the rocker panels and rear spoiler provide the final touches on overall balance of lift on the car. The optional carbon-fiber package provides enlarged splitter, hood air extractor, rear spoiler, rocker panels, and rear diffuser.

Those aren't the only aerodynamic devices. Under the car, the lower control arms at the rear are sheathed in plastic covers that Barrick says reduce drag by two counts.

Naturally, as a high-performance variant, once the necessary supporting infrastructure was established, the ATS-V needed a suitably more powerful powertrain.

**The ATS-V's twin-turbo 3.6-L V6 produces a seemingly effortless 464 hp.**

"Book matched" carbon fiber components save weight while looking good, as their grain is aligned in a mirror image like the words in an open book.

These relatively inexpensive stamped steel gussets strengthen the connection of the rear suspension to the unibody, contributing to the ATS-V's responsive handling.

It is an SAE-certified 464-hp (346-kW), 444-lb·ft (601-N·m) twin-turbocharged, direct-injected 3.8-L V6 engine matched to a six-speed manual or eight-speed automatic transmission. The car accelerates to 60 mph in 3.8 s and reaches a top speed of 189 mph (304 km/h).

The engine features turbocharges with low-ineria titanium-aluminide turbine and vacuum-actuated wastegates for reduced lag and stronger low-rpm response. The turbos' compressors are matched for peak efficiency at peak power levels, providing maximum on-track performance.

The engine's response is augmented by a patent-pending low-volume charge-cooling system that improves packaging efficiency while permitting maximum boost pressure. The engine itself enjoys some reduced inertia courtesy of its titanium connecting rods.

The Hydra-Matic 8L90 eight-speed automatic is the faster and more efficient transmission choice. The car features steering-wheel-mounted paddles for manual shift actuation, but when the ATS-V is switched to attack mode, the transmission's computer makes flawless shift decisions while the driver hammers the car around a circuit. In most respects, it is the better way.

However, for those drivers who think it is not the fun way, Cadillac offers the Tremec TR6060 six-speed manual transmission. We've seen automatic rev matching on downshifts elsewhere, including GM's own Chevrolet Corvette, but the ATS-V adds to that no-lift upshifts. The driver depresses the clutch, makes the manual gearchange using the H-pattern shifter, and releases the clutch pedal all as usual. But the accelerator pedal stays on the floor throughout and the car automatically cuts power just long enough to compete the shift motion.

Though this defies the accumulated muscle memory in anyone able to operate a manual transmission, it makes upshifts shockingly quickly. So, maybe computer-matched downshifts and automatic power-cut upshifts leave the driver doing less of the manual shifting job, both ensure that the shifts are smoother and faster.

The EPA fuel economy estimates for the automatic transmission-equipped ATS-V are 16 mpg city and 24 mpg highway, while the manual gets 17 mpg city and 23 mpg highway.

An electronically controlled limited-slip differential helps the power get to the road effectively, with a 3.73:1 final drive ratio for the manual transmission and a 2.85:1 for the automatic, thanks to its abundance of gear ratios that permit a very low first gear.

Apparently, enthusiasts will have to accustom themselves to both the notion of sporty automatic transmissions and sporty compact Cadillacs. ■

*Dan Carney*

# Cadillac's 2016 CT6 is a body-engineering benchmark

**While the CT6 launches with three gasoline engines, it will add four-cylinder and V6 diesels from GM's advanced light-vehicle diesel family, now under development, for Europe within three years.**

Cadillac's all-new 2016 CT6 full-size sedan is big for such a featherweight car. It weighs about 200 lb (91 kg) less than the BMW 5 Series but offers the interior volume of the short-wheelbase 7 Series, thanks to what is arguably the industry's most aggressive combination of lightweight materials, forming technologies, and new joining methods in a non-exotic sedan.

The CT6 weighs 3700 lb (1678 kg) in base-curb form with 265-hp (197 kW) 2.0-L turbocharged I4 and rear drive—5 lb (2.2 kg) lighter than the smaller CTS. Overall it is a technological tour de force, as competitors will ultimately discover in their first teardown analyses. Its top-line powertrain features a Cadillac-exclusive 3.0-L V-6, SAE rated at 400 hp (298 kW) and 400 lb·ft (542 N·m) and using the industry-first combination of twin turbochargers, direct injection, cylinder deactivation (General Motors' new system for OHC valvetrains), and stop-start.

Power is delivered to either the rear wheels or all wheels, depending on driveline configuration, via GM's new 8L45 eight-speed planetary automatic. An automatic decoupling AWD is standard on V6 cars. A new 3.6-L naturally aspirated V6 also is available, and GM's new family of light-vehicle diesels currently under development will provide four-cylinder and V6 engines for European CT6 models within three years, a GM executive said.

Inside the CT6, a "camera mirror" provides streaming video of 360° front and rear views while driving. GM's new third-generation night-vision system helps the driver identify people, creatures, or objects in the road ahead using thermal imaging, alerting the driver with a display on the cluster.

*Automotive Engineering* will publish detailed articles on the CT6's development and key technologies in the upcoming months.

The new Cadillac proves GM's commitment to "reinventing the large luxury sedan," as the brand's

President Johan de Nysschen said at the car's New York media launch. "Reinventing" translates into putting GM's flagship brand beyond par with the world's best in every comparative metric and keeping it there.

According to Cadillac Chief Engineer Dave Leone, and CT6 Chief Travis Hester, development began over four years ago with the aim to create a large (E-segment) rear-drive sedan with class-leading NVH performance offering the driving dynamics of a midsize. Included in the plan was rear-passenger room and interior execution featuring five-way-massaging front seats and reclining heated/ventilated rear seats also with massage. The front seatbacks have integrated and retracting 10-in infotainment screens.

Such lavish attention to comfort, convenience, and detail is expected by premium-vehicle customers in China, one of GM's production locations for CT6. U.S. production will begin at GM's Detroit-Hamtramck plant in late 3Q2015.

Lessons learned in body and suspension engineering of the C-segment ATS and D-segment CTS dovetailed with development of GM's new lightweight Omega vehicle architecture, first deployed on this model. 64% of the CT6 is aluminum, including all exterior panels. The multi-metals body-in-white "is lighter than that of the current Audi A6 and Mercedes E-Class and about 225-lb lighter than if we'd just used our CTS configuration and material set," Leone noted. The structure uses a carefully tailored mix of aluminum extrusions, 13 HPDC (high-pressure die castings), and high-strength and ultra-high-strength steel alloys (including the floorpan), for specific load-path and other applications.

Highlights of the HDPC components include a large die-cast front body hinge pillar, produced by Cosma. According to Hester, this component's size and complex geometry posed a challenge "but the Cosma guys really worked hard for us on this and came through," he explained.

The cast-aluminum hinge pillar "does the job of 35 parts in a traditional hinge-pillar application," Leone said, contributing to a body-structure that is "super stiff" while reducing the bill of material by 20%. An acoustic laminate panel supplied by Shiloh is used in the critical dash panel location to damp noise into the cabin.

To join the body structure, GM is employing its proprietary aluminum spot- and laser-welding processes—the structure contains 3000 spot welds—together with ample use of self-piercing rivets and "180 linear meters of adhesive bonding," Hester said.

The CT6 suspension uses a front high-arm short/long arm set up with aluminum control arms and knuckles. In the rear, there's a five-link, active-steering arrangement with multiple outer ball joints, with steel control arms and GM's MagneRide magnetorheological damping system available (not standard) on both ends. The system offers selectable drive modes including Tour, Sport, and Snow/Ice.

The new CT6 will ride on a choice of 18-, 19-, and 20-in wheels, with 255/45 tires. ■

*Lindsay Brooke*

# Cadillac CT6 to use advanced mixed-material and manufacturing approach

**Cadillac will use the auto industry's most comprehensive and advanced mixed-material manufacturing techniques to build its all-new CT6 top-of-range sedan. Material joining processes include aluminum laser welding, shown here.**

Cadillac will use the auto industry's most comprehensive and advanced mixed-material manufacturing techniques to build its all-new CT6 top-of-range sedan. Material joining processes include aluminum laser welding, shown here.

Cadillac's new CT6, set for release in the fourth quarter of 2015, will be built with various types of advanced and lightweight materials and manufactured using a combination of joining techniques, a first for any GM model. A $300 million investment in the Detroit-Hamtramck Assembly plant led to a new 138,000-ft2 (12,820-m2) body shop that includes 205 robots specifically dedicated to the manufacturing of the luxury sedan.

Among the techniques being used to join the CT6's dissimilar materials are aluminum spot welding technology; aluminum laser welding, which creates a seamless joining of exterior panels; self-piercing rivets, which are able to join different types of materials together with a clean appearance; flow drill screws, which are able to join different types of materials and are used in conjunction with adhesive; and aluminum arc welding and structural adhesive are also separately used for CT6 body assembly.

Twenty-eight robots weld both the inner and outer vehicle frames on the vehicle body in two separate framing stations, joining the body-in-white together from all angles. The robots are mounted above and beside the vehicle and can also reach beneath it. The two framing processes were choreographed to compensate for different microscopic vibrations.

Twenty-eight robots weld both the inner and outer vehicle frames on the vehicle body in two separate framing stations, joining the body-in-white together from all angles. The robots are mounted above and beside the vehicle and can also reach beneath it. The two framing processes were choreographed to compensate for different microscopic vibrations. ■

*Zachary Nocera*

# Cadillac pursues aluminum/steel mix for new CT6 luxury sedan

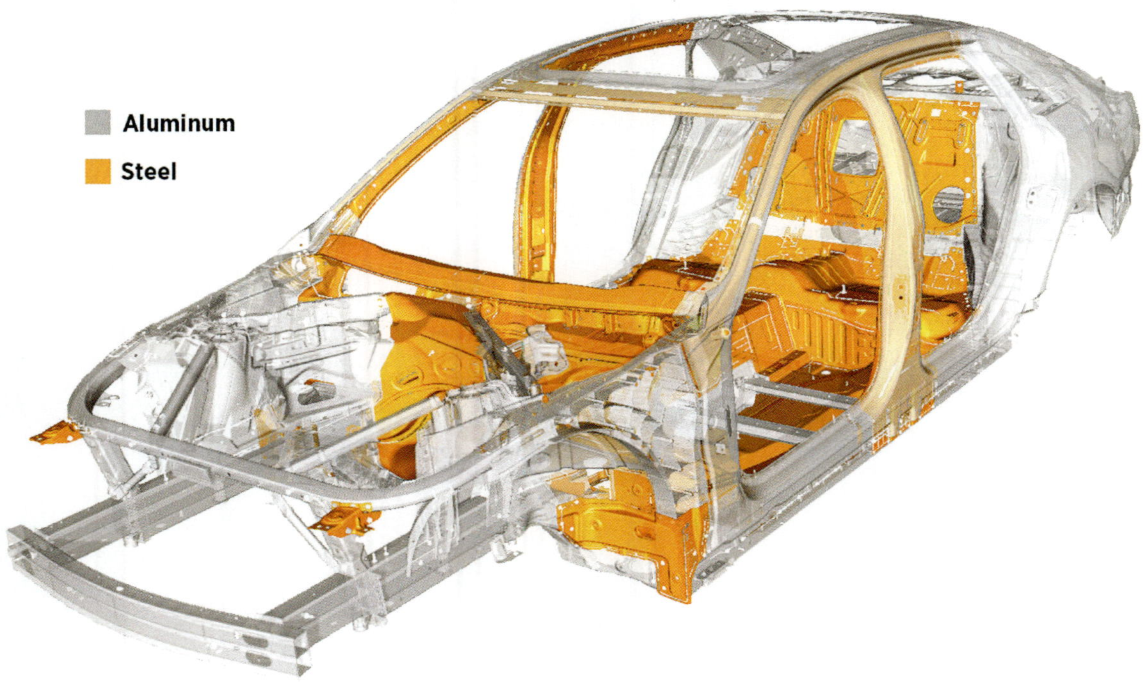


**The new Cadillac CT6 body is about two-thirds high-strength aluminum and one-third steel and high-strength steel.**

Next fall, Cadillac's new top-of-the-line model, the CT6 sedan, rolls out of General Motors' big Detroit-Hamtramck assembly plant and onto dealership sales floors nationwide to confront the likes of the Mercedes S-Class, Audi A8, and BMW 7 Series in the premium luxury large-car market.

To equip its flagship sedan for the challenge, Cadillac designers and engineers have built the CT6 on a unique vehicle architecture, an aluminum-intensive body structure that is fully one-third steel. Twenty-one patents are pending for the new full-size, rear-wheel-drive, transverse-engine configuration, a variation of GM's Omega II platform.

"The mix of materials helps save almost 200 lb compared to all-steel construction, and it's lighter than an aluminum design would be as well," said Travis Hester, Executive Chief Engineer. "We got the curb weight under 3700 lb, which is approximately the same as a CTS, but what's remarkable is that the CT6 is 8 in longer than the CTS," the Australian engineer stated with Down Under brio. "So the CT6 has about the same mass as a vehicle that's a full weight-class down in size."

## Cast alloy backbone

Shaving that weight helps maximize fuel economy, but it also enhances the sedan's driving dynamics: "We're lighter, quicker, yet we have great torsional rigidity to ensure world-class ride and handling performance, which was central to our goals. The structure delivers the chassis foundation we needed," Hester said.

**Conventional steel spot welding still takes a major role in the assembly of the new aluminum-intensive Omega platform-based sedan.**

**Cadillac uses new manufacturing processes to build the advanced, mixed-metals body structure of its new CT6 sedan.**

Thirteen complex high-pressure aluminum die-cast components comprise much of the lower structure of the CT6 body, he explained. "These linked node joints—the backbone—are huge castings, some of the biggest that our supplier is aware of. The large one in the tunnel area, for instance, gets us great powertrain response."

The Cadillac engineers employed hot-stamped, high-strength steel (HSS) strategically to reinforce the body structure and in conjunction with high-strength aluminum to create a tight, resilient safety-cage cabin structure whose underbody is lined with steel close-out panels to isolate it from noise and vibration. Aluminum sheet and extrusion components make up the engine compartment, trunk, roof, and rocker panels. Engineers added a high-strength aluminum impact bar to the rear of the vehicle, and a combination of high-strength alloy and HSS parts in the front and side impact zones to boost safety.

## Mixed materials allowed

"Four and a half years ago, when we began planning for a larger Cadillac, we did multiple investigations into materials selection," Hester recalled. At the time, the standard approach taken by much of the luxury car competition was to use aluminum alloy as the predominant structural material. "And we started going down that road for a little bit, but then we got away from it for a couple of reasons."

"Number one, our manufacturing guys had managed to come up with a series of technological enablers that allows us to construct joints between pretty much any material with any other material," the chief engineer said, stressing the significant amount of work GM's manufacturing research team had done to ready these new methods for the production line. "They did a lot of complicated R&D work that took a while."

As a result, "alloy to steel, casting to extrusion..., the flexibility in joining and fastening opened up combinations which we normally could not have put together," he continued. The free mixed-material design approach "let us use the best material in the best way in each location, integrating aluminum and steel where it makes sense."

The body engineers used a multi-disciplinary design optimization model, a CAE-based parametric simulation that finds the "best" way to meet weight and performance objectives. The engineers specified property ranges and potential materials: "Alloy grades, x to y, steel grades, x to y, and ran the simulation multiple times," he said. "The optimization model pretty much let every material select itself for every part. We continued to do that and it allowed us to really optimize every component on the car for its job."

**Each Cadillac CT6 body is rigidly joined by some 3000 aluminum spot welds and 180 m of structural adhesive bonding.**

**Cadillac knew that it needed high-technology engineering to compete in the premium luxury car market.**

One example is the lower structure of the cabin, which is constructed of steel close-out panels. The computer analysis highlighted that steel alone would be lighter than aluminum with the added weight of extensive sound-deadening material that would be needed to compensate in the aluminum design. "The most common comment we get is how quiet the car is," Hester noted.

"In another case, the design allowed us to combine what was over thirty parts into one part," he said. "In fact, the entire body structure has only 412 parts; that's 20% less than normal steel." Fewer parts takes out a lot of manufacturing variability and provides a remarkable improvement in precision and production efficiency.

After the mass optimization process is "when things started to get pretty exciting," he said. The large lightweight body made the car "be extremely wide and allowed us to add features that the competition does not offer such as the huge wheels—18-in to 20-in wheels." The twin-turbo V6-powered CT6, which debuts at the New York International Auto Show, has a stretched wheelbase and long rear doors and hood.

## Advanced assembly technology

To build the new body design, company management installed at the Detroit-Hamtramck plant a 138,000-ft$^2$ (12,820-m$^2$) body shop fully automated with some $384-million worth of high-tech joining and fastening tools and equipment, including 203 new assembly robots.

"For the Cadillac CT6 we have developed additional new body-construction techniques and technologies allowing various types of advanced and lightweight materials to be combined within the manufacturing environment like never before," said Cadillac President Johan de Nysschen when he announced the plant upgrade at the Washington Auto Show.

Several material-joining and fastening techniques are used to assemble the CT6, Hester said. "It starts off with 3000 alloy-to-alloy aluminum spot welds and includes conventional steel spot welds as well."

When joining alloy to steel, for example, and the design allows two-side access to the assembly equipment, he said, self-piercing rivets are typically used, especially when a clean appearance is needed. These sharp-tipped fasteners are applied at high pressure, then the end is folded over with a tool to form a mechanical interlock.

When equipment access is limited to one side and the joint is to be between differing part forms or materials, they typically use flow drill screws in conjunction with adhesive. Flow drill screws are fasteners that "self-extrude" as they pierce the metal being joined as the screw is rotated at high-speed. "The alloy castings have a big draw in many areas," he said, "so in that case we use flow drill screws." They are also used when attaching HSS components.

The new sedan has about 180 m (590 ft) of structural adhesive in it. Aluminum arc welding and aluminum laser welding, which creates a seamless bond between exterior panels, are also employed in the body assembly process.

On the CT6 production line, 28 robots descend on the vehicle body in two separate framing stations to weld both the inner and outer vehicle frames, simultaneously joining the body-in-white together from all angles. As the choreographed robot arms move in, out, and around the vehicle in concert, they resemble a symphony orchestra. ■

*Steven Ashley*

SAE Eye on Engineering Quick Take:

### Cadillac vs. Lincoln

Watch the video at: https://youtu.be/KzKcvY8wiKg

Cadillac vs. Lincoln is one of the world's great automotive rivalries. In this episode of SAE Eye on Engineering, Senior Editor Lindsay Brooke looks at the new Cadillac CT6 and Lincoln Continental. SAE Eye on Engineering can be viewed at https://youtu.be/KzKcvY8wiKg. It also airs in audio-only form Monday mornings on WJR 760 AM Detroit's Paul W. Smith Show.

Access archived episodes of SAE Eye on Engineering at www.sae.org/magazines/podcasts.

# Cadillac fortifies CTS for V duty

**Cadillac took its CTS-V sport sedan challenger to Road America to prove the car's on-track prowess.**

Cadillac continues its battle to achieve premium market esteem equivalent to that of segment leaders BMW and Mercedes-Benz by pushing for top quality and performance in each new model. The high-performance CTS-V is meant to put a stake in the ground, demonstrating Cadillac's seriousness and, to do that, the car is upgraded from the standard CTS in nearly every respect.

Visually, the changes start with the bodywork, which includes a lightweight carbon-fiber hood and revised front and rear fascias, fenders, rear spoiler, and rocker moldings. An optional carbon-fiber aero package adds a front splitter, enlarged wing, diffuser, and carbon-fiber air extractor on the hood. Enlarged intake openings and less obstructive grille mesh contribute to a 50% increase in airflow through the front of the car to its array of six heat exchangers.

"Every exterior detail earned its way onto the car for its aerodynamic function," stated Product Director Hampden Tener. Aerodynamic effectiveness is critical because the CTS-V is the fastest Cadillac ever; top speed is 200 mph (322 km/h) and it rockets from 0 to 60 mph (97 mph) in 3.7 s.

Those flared fenders spread to cover the 19 x 9.5 in front and 19 x 10 in rear wheels. These aluminum wheels are 45% stiffer than those used on the previous-generation CTS-V for more precise steering control.

They mount Michelin Pilot Super Sport summer-only performance tires: 265/35ZR19 in front and 295/30ZR19 in the rear. As with the ATS-V, these custom-tuned Michelins feature three distinct rubber compounds in the tread, with a hard outer shoulder, grippy inner third, and a rain-optimized center section.

Inside the wheels are Brembo brakes with staggered six-piston front calipers and four-piston rears. The rotors are a two-piece design, with a lightweight aluminum center section and cast iron friction surface. The 390 mm (15.4 in) diameter fronts are 365 mm (14.4 in) rears are treated with a ferritic nitro carburized process for corrosion resistance. Customers can choose among three available matte-finish colors for the calipers.

At the rear, the CTS continues with a five-link independent suspension arrangement, while the front McPherson strut setup enjoys dual lower ball joints and a direct-acting anti-sway bar for more precise control. These zero-compliance ball joints, stiffer springs, and a stiffer anti-roll bar combine to provide a 20% increase in roll stiffness compared to the traditional elastomeric bushings.

Wheel motion is controlled by BWI Group magnetorheological shock absorbers, the one-time MagnaRide technology, now in its third generation. A 40% faster response time than before lets the CTS-V respond to changes in the road surface every inch the car travels at 60 mph.

A ZF Steering Systems Servotronic II variable-ratio electric power steering rack has 14% greater system stiffness for improved steering feedback and precision.

Underpinning all of these moving parts is a chassis that is reinforced for a 25% increase in stiffness compared to regular CTS models. Upgrades include a shock tower-to-tower brace, a shock

**Lighter rotors let the smaller-displacement Eaton supercharger spin up to boost-producing speed more easily.**

tower-to-plenum brace, V brace for the engine compartment, strengthened rocker bulkhead, stronger braces between the rear suspension cradle and the chassis rocker area, and braces from the upper tie bar to the rear bumper. Additionally, a bolt on 25 mm (1 in) thick aluminum panel connects the front cradle to the chassis more securely.

With all this reinforcement, bigger brakes, and enlarged heat exchangers, the CTS-V's 4145-lb (1880-kg) curb weight comes in 100 lb (45 kg) lighter than that of the previous car.

The CTS-V's powerplant is General Motors' supercharged overhead valve 6.2-L aluminum block LT4 V8 producing an SAE-certified 640 hp (477 kW) at 6400 rpm and 630 lb·ft (854 N·m) of torque at 3600 rpm. Cadillac notes that these numbers top those of rivals like the 577-hp (430-kW) Mercedes-Benz E63 AMG or the 560-hp (418-kW) BMW M5. Cadillac predicts the CTS-V will achieve U.S. EPA fuel economy scores of 14 mpg city and 21 mpg highway.

As with the current-generation Chevrolet Corvette Z06, which debuted the LT4 engine, the CTS-V's powerplant features a smaller-displacement version of the Eaton supercharger, whose smaller rotors' reduced inertia spin up more quickly under acceleration. The cylinder heads are Rotocast A35T6 aluminum for its greater strength and tolerance of heat.

Power transfers from the LT4 to GM's Hydra-Matic 8L90 planetary automatic transmission through a reduced-diameter torque converter. Now 258 vs. 300 mm (10.2 vs. 11.8 in) on the previous car, this torque converter's reduced inertia contributes to the transmission's ability to execute upshifts in just 150 ms, putting it in the hunt with dual-clutch transmissions for rapid shift times while providing the smooth low-speed driving characteristics expected of automatic transmissions.

The transmission's shift characteristics, along with control of the magnetic ride control shocks, power steering assist, throttle response, and many other variables, are determined by the drive mode selector. In Touring mode shifts are optimized for smoothness and fuel efficiency, Sport mode prioritizes keeping the engine revving, and Track mode delivers an amazingly correct shift schedule for all-out racing. There's also a snow mode that is probably better not tested on the standard summer-only Michelins.

The rear differential is computer controlled so it can be totally open or provide peak pre-load torque, if needed, to transfer power to the tire with grip.

Early CTS-Vs suffered axle hop under acceleration, and Cadillac's solution is to use different diameter axle shafts so that they operate with different resonant frequencies. In this case, the driver's side axle is 52 mm (2.0 in) in diameter compared to 28 mm (1.1 in) for the passenger's side.

The car's fuel tank is also located in that vicinity, and Chief Engineer Dave Leone took the opportunity while pointing out critical bits on a rotisserie-mounted car to mention that prototypes had suffered from fuel starvation at low fuel levels. His team diagnosed the problem by installing a clear gas tank for testing that let them see what was happening inside. Now, owners can hammer the CTS-V around a track at high levels of lateral acceleration (we saw 1.25 g in corners at Road America) with the fuel as low as an eighth of a tank without suffering any fuel-pickup problems, Leone said.

Inside, the CTS-V is fitted with extra leather and unique optional $2600 16-way adjustable Recaro front bucket seats. The available Webasto panoramic sunroof, whose large opening easily lights front and back rows, lend the cabin an airy atmosphere.

As the first 2016 Cadillac to reach showrooms, it also debuts the second generation of the company's signature CUE touch-screen infotainment and HVAC control interface.

Addressing complaints about the slow, balky performance of the first-generation system, Cadillac has upgraded CUE computing hardware to the Freescale Semiconductor iMX6 processor that, at 4000 million instructions per second, is three-times faster than the original chip. Response time is down to 100 ms, which is within shouting distance of an Apple iPhone's 80-ms response time, so users will find it satisfactory, reported Leone.

Critically, destination routing in the GPS navigation system is much faster than before, eliminating a key headache. Further, programmers improved the navigation system interface with the simple inclusion of a "cancel" button on the map while the system provides guidance. Previously, drivers had to navigate through three layers of menus to perform this common task.

**The optional carbon-fiber package adds a downforce-inducing spoiler for critical stability at 200 mph.**

**Massive six-piston Brembo front brake calipers are available in three festive matte colors.**

The center stack display also can now mirror iPhone apps, giving drivers somewhat less distracting access to a few key apps that are enabled for use in the car. Google's Android Auto is also in the works, according to Leone.

The CTS-V also features the built-in video camera and Cosworth data-acquisition system seen previously on the Corvette Z06 and Cadillac ATS-V, letting drivers shoot video of track laps while recording data from those laps that is overlayed on the video for easy viewing. With the Cosworth Toolbox application on a computer, CTS-V drivers can drill down into their data and compare laps or even against another drivers data to learn speed secrets for faster laps.

Or they can use the data to prove to their friends that the CTS-V is indeed faster than its German competitors. ■

*Dan Carney*

# 2016 Camaro: GM unleashes its lighter, faster Mustang competitor

**GM held the world debut of the 2016 Camaro on Detroit's Belle Isle Park and included over 1000 Camaro owners from all over the U.S. and Canada, and a fantastic Camaro Museum on site.**

During the early phase of 2016 Chevrolet Camaro development, Chief Engineer Al Oppenheiser one day found himself in a meeting with a group of GM Powertrain engineers he'd never met before.

"After we all said hello, my boss turned to me and said, 'Al, these are the four-cylinder turbo guys. You should get to know them.' And I said to myself, well, then I guess we need to get acquainted," Oppenheiser recalled.

Many months later on a prototype drive in Germany, Oppenheiser—a GM Performance veteran who admits to a love for muscular small-block V8s—had his first chance to wring out a Camaro mule powered by the newest and most non-traditional-Camaro engine—GM's turbocharged 2.0 L rated at an SAE-certified 275 hp (205 kW) and 297 lb·ft (402 N·m).

"I had intended to take the car for a short drive on local public roads and ended up returning five hours later," he told *Automotive Engineering*. "The car with the turbo engine is simply a blast to drive. Light and nimble. Felt lively, nice crisp turn-in. I was delighted," he said.

There also is ample customer delight in the development story of the all-new, sixth-generation Camaro that launches this fall in LT and SS trim. A convertible is officially in the works, and mention of a Ford Shelby GT350-gobbling Z/28 version elicits grins and "no comment" from engineers. The biggest news is the platform: with their primary mission being to significantly improve Camaro's dynamic performance, acceleration, and fuel economy, Oppenheiser's young development team looked immediately to significantly reducing mass and improving structural stiffness (by 28%).

To achieve this they moved Chevy's iconic rear-drive sporty coupe from its GM Australia-derived Zeta architecture, to the lighter mixed-materials Alpha platform that debuted on the 2013 Cadillac ATS and is also used on the CTS. And they created a more compact Camaro—at 188.3 in (4783 mm) the 2016 model is 2.3 in (58 mm) shorter overall, and rides on a 1.6 in (41 mm) shorter wheelbase at 110.7 in (2812 mm). Combined with the new platform's inherent mass efficiency, the slightly smaller '16 car is 200 lb (91 kg) lighter, on average across the model range. Comparing 2016 to 2015 base models, the new version is nearly 300 lb (136 kg) lighter.

The base turbo four that initially surprised Oppenheiser is claimed to accelerate from rest to 60 mph (97 km/h) in less than 6 s while delivering more than 30 mpg highway fuel efficiency—making it the most fuel-efficient Camaro ever. The car's six-combination powertrain lineup also includes a

**Camaro Chief Engineer Al Oppenheiser is a veteran of GM's Performance Vehicle group who was impressed the first time he drove a 2.0-L turbocharged four-cylinder mule.**

335-hp (250-kW) 3.6-L V6 and a 455-hp (339-kW) 6.2-L LT1 V8. All are direct injected and their output SAE-certified. The 2.0-L and 3.6-L engines are mated with a choice of Tremec TR316 six-speed manual or GM's new 8L45 eight-speed automatic transmissions.

The V8 powering the SS model produces 455 lb·ft (617 N·m) and employs the Tremec TR6060 six-speed manual and 8L90 eight-speed automatic. The TR6060 unit is equipped with active rev matching, a driver-selectable feature that automatically blips the throttle (using signals from a Hall Effect sensor on the shift linkage) for precision downshifts.

Other key technologies aimed at tailoring Camaro's dynamic performance to the driver's demands include BWI-sourced magnetic ride control suspension dampers (a first use on Camaro SS) and ZF's rack-mounted electric power steering that GM performance-car engineers universally love while quietly bemoaning its price premium.

There are also four driver-select operating modes—snow/ice, tour, sport, and track (the latter on SS models only), and a dual-mode exhaust system that uses new electronically controlled valves to let the exhaust bypass the mufflers under moderate-to-wide throttle openings. At its most restrictive setting, the system is sneak-home-at-2 a.m. quiet. Conversely at WOT it plays a raucous back-straight symphony.

How effective are the performance improvements? According to GM engineers, the lap times of the base '16 Camaro SS are faster than those achieved by the '15 Camaro 1LE package, a track-day niche model that features suspension bits from the Z/28 bill of materials.

Inside Camaro, the cabin design focuses on improved human-machine interface and driver/passenger ergonomics. Tactile surfaces have a more premium look and feel than those of the Zeta-based car, and a "spectrum lighting" feature illuminates the IP, door panels, and center console in a 24-effects light show.

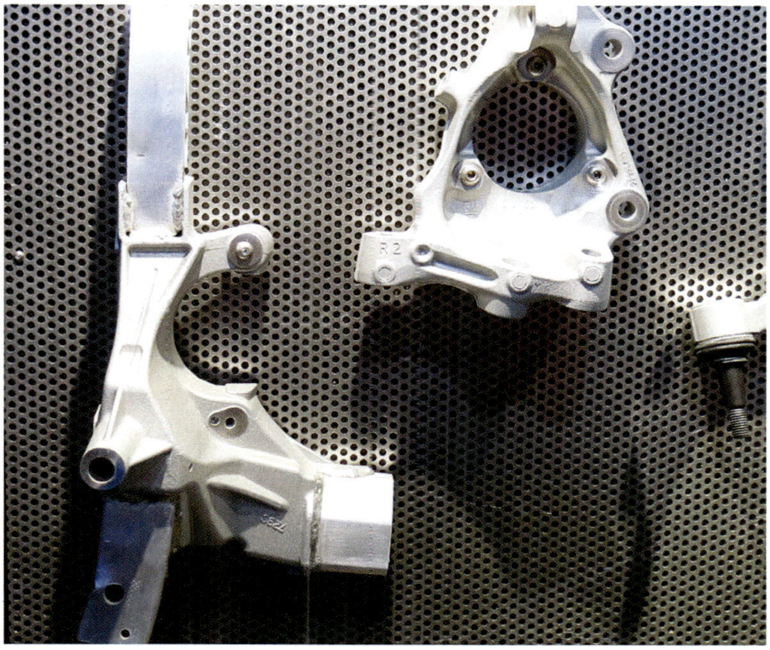

**The 2016 Camaro moves to aluminum-intensive suspensions front and rear, including a double-pivot front arrangement that is designed to optimize steering feel and response with the ZF electric power steering.**

**Packaging the mighty 6.2-L LT1 V8 in the new Alpha-based engine bay required some modifications, including a 10-qt wet sump oil pan.**

## Iterating Alpha for Camaro duty

Moving Camaro to Alpha wasn't entirely part of the original product plan for GM's latest architecture, noted Mark Reuss, GM Executive Vice President for Global Product Development.

"About a year and a half ago, we took the Zeta-based car and looked at the proportion models on the patio really, really closely before we decided on Alpha for this," he explained. "The capability to do a Camaro was always designed into Alpha, but it wasn't decided then. You've got to let the market mature a bit, to see where others are headed. There were some things we had to do [to the structure] to get the Camaro proportion."

Various design iterations followed, he said. The front structure is unique to Camaro. It is longer than that of ATS to create the required dash-to-axle ratio for the "Camaro look" and is widened to provide the desired 62.5-in (1588-mm) front track width on LT, and 63 in (1600-mm) on SS. In addition, 20% of the V8 engine has been modified to package it in Camaro, including new fabricated tubular steel "tri-Y"-style exhaust manifolds.

The crosscar beam is now an aluminum extrusion that saves 9.7 lb (4.4 kg) versus the steel beam on the current-generation Camaro. Front MacPherson strut and five-link independent rear suspension assemblies also are aluminum-intensive and account for a 21% mass savings compared with the 2015 car. Illustrating GM engineers' attention to detail in slashing excess grams, suspension links on some models employ a structurally optimized design made with a rigid thermoset composite material that's lighter than aluminum.

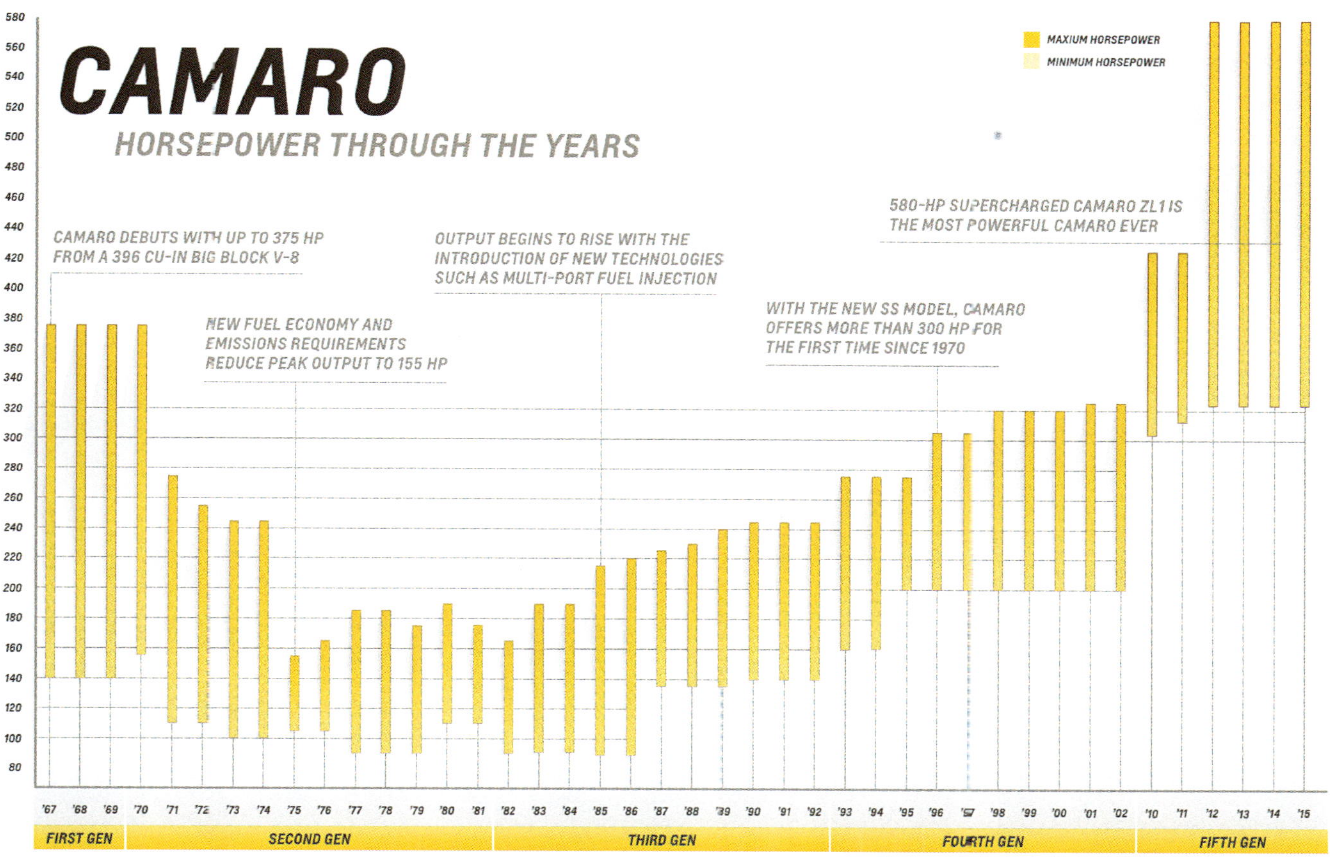


"We already had a good IRS with Zeta, so we knew we'd have a capable track car," Reuss observed. "But how do you do that and take the next step and take mass out of the car without ruining the value equation? Well, the only exterior panel that's aluminum on this car is the hood. But look underneath it and see all the things we did, piece by piece. The lightening holes in the steel rear suspension links show our philosophy.

"We knew if we did Camaro off of Alpha we'd get all the things Alpha brings to it—including [production] scale," he noted. GM will build the Camaro at its Lansing (MI) Grand River assembly complex, moving the nameplate from its former build at Oshawa, Ont.

## A 50-year battle

The current Camaro's sales success has been a surprise within the industry. GM put the nameplate on production hiatus for seven model years after 2002, returning in 2010 with the Zeta-based model that replaced the ancient F-car. Almost immediately Camaro outsold its old nemesis, the Ford Mustang, ironically the original "pony car" that inspired GM to create Camaro for the 1967 model year.

For six straight years the Zeta-based Camaro put Mustang on the trailer, sales-wise, reaching a peak of over 88,000 sales in 2011. Then late last year, armed with an all-new Mustang, Ford finally eclipsed Camaro's sales by a wide margin. With both classic American sports/muscle machines now on roughly the same development cadence, their ongoing technology, performance, and sales battle that is approaching 50 years old will be interesting to watch. ■

*Lindsay Brooke*

SAE Eye on Engineering Quick Take:

### Camaro convertible top

Watch the video at https://youtu.be/tPfPEpfDILA

What would the Chevy Camaro be without a convertible version? In this episode of SAE Eye on Engineering, Senior Editor Lindsay Brooke looks at what GM engineers claim is the most advanced top system in the sporty-car segment. SAE Eye on Engineering can be viewed at https://youtu.be/tPfPEpfDILA. It also airs in audio-only form Monday mornings on WJR 760 AM Detroit's Paul W. Smith Show.

Access archived episodes of SAE Eye on Engineering at www.sae.org/magazines/podcasts.

# 2016 Chevrolet Cruze is larger, longer, and sheds 250 lb

**Overall height on the 2016 Cruze has been reduced by nearly an inch versus the 2015 car, which helps enable a 0.29 Cd.**

General Motors' best-selling global product, the Chevrolet Cruze, gains an all-new platform for 2016 that helps make the C-segment model longer, roomier, more aerodynamic, and lighter—shedding 250 lb (113 kg)—while gaining the new 1.4-L direct-injected turbocharged gasoline engine with stop-start and a suite of available connected-vehicle technologies.

**The 7-in MyLink screen in the IP is compatible with Google Android Auto and Apple CarPlay—the latter available with an 8-in screen, and offers wireless phone charging.**

A 1.6-L turbodiesel sourced from GM's Szentgotthárd, Hungary, plant currently offered in Opel models is scheduled to debut in North American Cruze models in early 2017.

To be sold in over 40 countries, the new Cruze shifts to GM's D2XX compact platform which succeeds the current Delta II that is shared with Buick Verano, Chevy Equinox, Opel Astra, and others. The four-door notchback body style is expected to be joined by a five-door hatch. A slight increase in wheelbase, to 106.3 in (2700 mm) from the current 105.7 in (2685 mm), along with 2.7 in (69 mm) longer overall length, makes the new Cruze larger than competitors Honda Civic, Ford Focus, Dodge Dart, Mazda3, Hyundai Elantra, Nissan Sentra, and Volkswagen Jetta.

Inside, the car is roomier than its predecessor and boasts more rear legroom—36.1 in (917 mm)—and 2 in (51 mm) more rear knee room than Focus and Elantra, according to GM. Front and rear tracks carry over from the current Cruze, at 60.8 in (1544 mm) and 61.3 in (1557 mm), respectively.

Like the 2016 Chevrolet Volt, the new Cruze's overall height has been reduced—by nearly 1 in (25 mm), giving it a sleeker look and enabling a 0.29 coefficient of drag (Cd)—a factor in the car achieving the estimated 40-mpg highway fuel efficiency, engineers noted. EPA fuel efficiency testing was not yet completed at the car's June 24 reveal.

The new body architecture, claimed to be 27% stiffer overall than the 2015 model, is expected to help further decrease cabin noise as well as improve vehicle dynamic performance and occupant safety.

**GM's new 1.6-L Ecotec turbo gasoline engine switches to a die-cast aluminum cylinder block and direct fuel injection, gaining 14 hp (10 kW) versus the incumbent iron-block 1.4-L turbo. Engineers focused on NVH reduction and "PFI-like" injector noise from the 2300-psi (15.9-MPa) system. The engine has 10:1 compression and runs on Regular octane fuel.**

The body shell uses about 8% hot-stamped/high-strength steel alloys that help reduce mass by 53 lb (24 kg) versus the incumbent Cruze's.

Executive Chief Engineer Ron Arnesen calls the Cruze's chassis system "perhaps the most mass-efficient in the segment," with aluminum and optimized steel components used in the MacPherson strut front suspension and in the torsion-beam rear setup. Wheel size ranges from 15 to 18 in.

GM's switch to its new-generation 1.4-L turbocharged engine for Cruze contributes to the car's interior quiet. Measured noise levels are up to 50% quieter—6 dB(A)—than the benchmark VW 1.4-L TSI, Small Ecotec Gas Engines Chief Engineer Tom Sutter told *Automotive Engineering*. The engine's open-deck, diecast aluminum cylinder block features a bedplate architecture that stiffens up the lower end, also contributing to inherently better NVH attenuation.

The aluminum block is also 20 lb (9 kg) lighter than the grey iron block in the current-gen Family Zero 1.4-L turbo engine. Still a long-stroke design, the new engine also gains direct injection that, along with a 10:1 compression ratio, helps boost its SAE-certified power rating to 153 hp (114 kW) at 5600 rpm—up from 138 hp (103 kW) on the previous port-injected unit. SAE rated torque is 177 lb·ft (240 N·m). Low-friction piston rings, camshaft drive, and oil pump are used. Transaxles are a standard M32 six-speed manual and available 6T35 six-speed automatic.

Inside Cruze, customers will find the Chevy-signature dual cockpit layout with vast safety capability and infotainment connectivity. The 7-in MyLink screen in the center stack is compatible with Google Android Auto and Apple CarPlay—the latter available with an 8-in screen, and offers wireless phone charging. Android Auto requires a phone running the Android Lollipop 5.0 operating system or above, while Apple CarPlay requires an iPhone 5 or later. There's standard OnStar 4G LTE with Wi-Fi hotspot. Available safety features include side blind-zone alert, rear cross-traffic alert, lane-keep assist, and forward collision alert.

Cruzes sold in North America are being built at GM's Lordstown, OH, complex, which received a $50-million upgrade for the 2016 program. ■

*Lindsay Brooke*

**Mark Reuss, GM Executive VP Global Product Development, unveils the 2016 Chevrolet Cruze C-segment sedan June 24, 2015 in Detroit.**

# 2016 Malibu sheds 300 lb, adds new hybrid system

**Malibu's development team stretched the wheelbase to increase rear passenger comfort. Claimed Cd is 0.29.**

Entering its ninth generation, the 2016 Chevrolet Malibu zoomed through its development cycle faster than any General Motors vehicle in recent memory. It gains a new platform and powertrains—including an all-new dedicated strong-hybrid model with some EV-only capability—aimed at putting the nameplate at the head of North America's most competitive segment.

"This was a fast program—about three years," Chief Engineer Jesse Ortega told *Automotive Engineering*. "We've got at last count 17 competitors for Malibu, in a hot segment that represents 14% of the industry." The car's design was the province of the team, under Chevrolet Executive Design Director John Cafaro, that delivered the current Impala, while the car's new Epsilon II architecture has its roots in Russelsheim.

The Malibu Hybrid model shares elements of its electrified-drive system with the 2016 Volt. It carries a U.S. EPA rating of 47 mpg combined (48 mpg city/45 mpg highway). Its numbers exceed those of hybrid versions of the 2015 Ford Fusion, Hyundai Sonata, Kia Optima, and Toyota Camry, and equal to that of the Honda Accord Hybrid. Ortega said the strong-hybrid drivetrain was designed into the vehicle structure from the onset.

Enabling "great interior and exterior styling, with best-in-class interior comfort," particularly more—1.5 in (38 mm) rear leg room, were the program's main targets, Ortega said. The cramped rear seat area was the main complaint of current Malibu owners, GM engineers said. It was one reason why sales fell 6% from those of the seventh-generation car.

## Target: 3375 pounds

The new Malibu's fastback, six-window greenhouse "makes a contemporary statement," noted Jaymer Starbody, Lead Exterior Designer. He explained that the new model's bolder proportions are the result of "pushing tires and sheetmetal out to the corners," lowering the shock towers within the body structure to reduce cowl height by 19 mm (0.74 in), and stretching the wheelbase 3.6 in (91 mm) and overall length by 2.3 in (58 mm). The car's 73-in (1854-mm) overall width is carried over from the previous generation Malibu.

Starbody claims a 0.29 Cd for the new Malibu's aerodynamics, aided by active grille shutters on LS and LT trim levels.

At the top of Ortega's must-do list was delivering "significant" mass reduction with the larger car—and to make Malibu the lightest hybrid model in the midsize segment. "Our target was to get it within the 3375-lb (1530-kg) test weight class, which we did, while achieving NHTSA 5-star (crash ratings)," he said.

Ortega describes his body engineers' FEA focus as "intense," which ultimately led to a 300-lb (136-kg) overall curb weight reduction versus the 2015 model. About 120 lb (54 kg) of that—more than 33% of the total weight savings—comes from the switch to the more mass-efficient platform. Epsilon II brings a higher percentage of high- and ultra-high-strength steels with reduced wall thickness on some components. The Malibu's clamshell hood is its only major exterior closure in aluminum alloy and saves 5.6 lb (2.5 kg) versus the previous steel hood.

According to GM structural engineer Amir Sewani, the FEA work also resulted in significant parts consolidation and reduced bill of materials.

Inside the body shell, Ortega's team hunted for excess ounces throughout the interior. The optional leather seats are collectively 40.6 lb (18.4 kg) lighter than those in the current model. About 1.5 lb (0.7 kg) was saved with new trunk-liner material and another 1.1 lb (0.5 kg) was lost with new, proprietary interior carpet material that also is claimed to be more durable than that used previously.

A new center console uses a one-piece structural shell that sheds 1.75 lb (0.8 kg) over the incumbent Malibu. New power window regulators collectively save 4.2 lb (1.9 kg)—a 34% reduction over the 2015 car. Another 2.2 lb (1 kg) was eliminated by simplifying the IP and its cross-car structure. The new center stack is a cleaner, simplified design that makes it easier to read and use the HVAC and infotainment controls. There's also space to store mobile devices and inductive charging for phones on some trim levels.

## Hybrid platform engineering

The biggest news on the propulsion front is Malibu's all-new, GM developed two-motor hybrid drive system that enables some EV-only driving. It also marks Chevrolet's first application of exhaust gas heat recovery (EGHR) that uses exhaust heat from the new 1.8-L, long-stroke, direct-injection four-cylinder ICE (internal combustion engine) to warm the engine and cabin at cold starts.

"The new hybrid drive is an example of 'platform engineering' with this and our new Voltec-2 unit used in the 2016 Volt," noted Darryl Wilson, the Hybrid Chief Engineer. He said the two systems were developed in parallel and share their DC inverter and power electronics, including Delphi Viper IGBTs.

The drive unit uses two electric machines coupled to a planetary gearset to give "the functionality of a CVT," Wilson said, "but it won't feel like a CVT—or like a Volt." The second motor enables downspeeding of the combustion engine and allows it to run close to or at its peak BSFC (brake specific fuel consumption).

A 1.5-kW·h battery pack, its 80 lithium-ion cells supplied by Hitachi, provides electric power and enables EV-only propulsion at up to 55 mph (89 km/h). The ICE automatically engages at higher speeds and loads when necessary to provide additional power. Total system power is rated at 182 hp (136 kW).

There are also two other all-new Ecotec engines for the non-hybrid models. The standard long-stroke 1.5-L turbo inline four from GM's new Small Gas Engine family is fitted with stop-start. It is 86 lb (39 kg) lighter than the outgoing model's standard 2.5-L unit and is claimed to deliver 160 hp (119 kW) at 5600 rpm and 184 lb·ft (250 N·m) at 2000-4000 rpm. (All outputs are estimated, pending completion

**GM's new strong-hybrid system that launches on the 2016 Malibu features two GM-designed electric machines and power electronics shared with the second-gen Volt.**

**Malibu prototype in a test chamber: Reducing cabin NVH from the already low levels of the previous model was a program bogey.**

of SAE testing when this article was written.) The base engine delivers an estimated 37 mpg highway with the 6T45 six-speed automatic. An optional 2.0-L turbo with square 86-mm (3.46-in) bore and stroke, is estimated to deliver 250 hp (186 kW) at 5300 rpm and 258 lb·ft (350 N·m) at 1700 rpm. The 2.0 L will power the Aisin-designed AWF8F45—GM's first eight-speed automatic for FWD applications.

In the realm of connectivity, infotainment, and electronic safety, Malibu's suite of new technology features includes available OnStar 4G LTE, and Chevrolet MyLink with a 7-in or 8-in diagonal color touch screens. There's also an all-new Teen Driver feature, which allows parents to view their kids' driving statistics such as maximum speed and warning alerts.

New available active safety features include a front pedestrian alert that uses a forward-looking camera and has brake intervention; lane-keep assist with lane departure warning; front and rear park assist; and adaptive cruise with front automatic braking. ■

*Lindsay Brooke*

# 2016 Volt: GM gives its PHEV longer range, less weight, easier charging

**The 2016 Volt shows off its classy new sheetmetal including an aluminum hood. Though the car is primarily a "bottom feeder," it uses active grill shutter for improved aerodynamics.**

The world's best-selling plug-in hybrid vehicle has been re-engineered from a clean sheet, on a new vehicle architecture, giving the 2016 Chevrolet Volt 30% greater electric vehicle (EV) range—estimated 50 mi (80 km) total on the battery alone—and more than 100 mi (160 km) additional range using both the new 18.4 kW·h battery and the equally new 1.5-L range-extending gasoline internal combustion engine (ICE).

General Motors engineers estimate the new Volt can travel 420 mi (675 km) using its full battery state of charge and the contents of its 8.9-gal (33.7-L) gasoline tank. Fuel economy is expected to be 41 mpg combined and 102 mpg-e when EPA testing concludes. Given that Volt owners overwhelmingly charge their cars at least twice a day during typical operation, engineers reckon the new vehicle should deliver an average of 1000 mi (1610 km) of driving per each tank of gasoline.

## Drops nearly two EPA weight classes

"Range was the most-requested improvement our customer base of over 73,000 owners asked for in the new Volt," explained the car's Executive Chief Engineer (and veteran of the original Volt program), Pam Fletcher. "This is GM's most studied group of owners, even more than Corvette. We used the OnStar feature in collecting user data and found they also wanted more 'fun-to-drive'—and they wanted a good-looking rather than quirky exterior."

The Volt development team, with the full support of GM top management in this key program, delivered those customer wants and more. As the photo at the top of this article shows, the Gen-2 Volt "isn't a funny-looking/cute car anymore," observed John Cafaro, Executive Director of Chevrolet's global design office. It's now got the handsome (to the author's eyes) "projectile look" front end, now with active aero shutters in the grill, that's become a strong styling cue across the Chevy range—not surprising since the Cafaro-led team that penned the Impala also designed the new Volt.

Along with the greater driving range comes a claimed 19% percent improvement in EV mode acceleration from zero to 30 mph (48 km/h), in 2.6 s, and a 7% improvement from zero to 60 mph (97 km/h), in 8.4 s. Some of this performance boost is due to significant mass-reduction actions. Overall, the 2016 Volt drops nearly two EPA weight classes [a class is 125 lb (57 kg)]—a feat that's nearly unheard of in most new vehicles where simply achieving mass parity from one generation to the next is considered a win. The new Volt is in the 3875-lb (1757-kg) weight class.

"Well, admittedly the first-gen car was a bit overdesigned, making it a bit overweight, but we were counting ounces on every [second-gen] component," said Andrew Farah, the Chief Engineer on both new and old Volts.

**Volt's redesigned interior is a dual-cockpit design featuring two 8-in screens. GM will debut its next-generation infotainment system at the NY Auto Show.**

The switch to GM's new more mass-efficient Gen-2 Delta platform (Volt is the first application) with its high percentage of high-strength steel alloys helped the cause, as did switching to aluminum hood, liftgate, and cylinder block. The new Voltec drive unit sheds 100 lb (45 kg) versus its predecessor, and the new battery pack of 360-V nominal/400-V peak voltage is about 21 lb (9.8 kg) lighter, too. The new Volt also benefits from further aerodynamic improvements in the upper- and under-body areas.

Back-seat passengers may appreciate that the seat now accommodates three persons, though to get into the additional perch still requires being smallish in dimension and scaling the rather tall center tunnel under which the large T-shaped lithium-ion battery pack resides. The pack itself features revised Li-ion cell chemistry; the cells remain a pouch-type form factor and are still supplied by LG Chem from its Holland, MI, plant.

The car's two-motor Voltec drive system operates both motors together in more driving situations in both EV and extended-range operation. The ability to use both motors contributes to the car's improved acceleration performance, Farah said. For a comprehensive look at the Gen-2 Volt propulsion system and battery published by *Automotive Engineering* last October, go to: http://articles.sae.org/13666/

## Charging improvements

Current Volt owners provided a wealth of data that helped the development team improve the new Volt's charging system and charging experience. Owners now can set their charging preferences exclusively for "home" charging, and the vehicle will automatically adjust to that setting when it is at "home" based on GPS data," explained Farah. More than half of Volt owners charge on 110 V, Fletcher noted, and the car is equipped with an SAE J1772 "combo connector," so Level 3 capability wasn't in the product plan for 2016, she said.

Owners can pre-set their Level 1 (120 V) charging level at 8 or 12 A and can charge immediately, set a departure time for each day of the week, or set a departure time and a utility rate schedule to charge only at off-peak rates. These settings only have to be programmed once and the Volt will default to them when the vehicle returns to its home location.

There's also a new charging-status system that features a specially designed tone that indicates when charging has begun, with additional tones for delayed charging. It will also indicate if the charge port door on the left front fender was left open after the last unplugging.

A new 120-V portable cord set has a simpler, more compact design and its storage bin was relocated to the left side of the Volt's rear cargo area, above the load floor, for improved accessibility.

The Volt's standard and available safety systems suite features a standard rear-vision camera, 10 standard air bags, and available lane keep assist with lane departure warning; side blind zone alert with lane change alert; rear cross traffic alert; forward collision alert with following distance indicator; front automatic braking, and advanced park assist with front and rear park assist (semi-automatic parallel parking). The rear parking brake is a new "motor-on-caliper" design, Farah said.

Start of production is slated for early 3Q15 at GM's Hamtramck, MI, assembly plant. The Volt's battery pack is built at GM's Brownstown, MI, plant. The new Voltec drive unit is manufactured at GM's Warren, MI, Powertrain plant. The 1.5-L engine will be manufactured at GM's Toluca, Mexico, engine plant for the first year of production, then shift to the Flint, MI, engine plant. Approximately 70% of the car's content is U.S. and Canadian built, a nearly 20% increase from the first-gen car. ■

*Lindsay Brooke*

# GM unveils more efficient 2016 Volt powertrain

**Electro-mechanical heart of the Gen-2 Volt powertrain is the all-new Voltec electric transaxle, shown here in sectioned view with direct-mounted TPIM. Note twin electric machines.**

"I can't think of a powertrain we've re-engineered more extensively within a five-year period than this one," observed Larry Nitz, GM's Executive Director, Transmission and Electrification, speaking about the range-extender hybrid system in the second-generation 2016 Chevrolet Volt. Major changes to the car's battery, Voltec electric drive system, and gasoline-engine generator—all are essentially new—are expected to increase its overall driving range and energy use by up to 12% compared with the current Volt, he said. The 2015 Volt's driving range is rated at 380 mi (611 km) after depleting both the 16-kW·h battery and 9.3-gal (32-L) gasoline tank.

Nitz and key Volt program engineers unveiled the new powertrain to *Automotive Engineering* and other media at GM's Warren, MI, transmission plant, where the all-new 5ET50 electric drive unit will be manufactured. The 2016 Volt will debut at the North American International Auto Show in January; SOP is scheduled for late summer 2015. At the Detroit show, GM will announce range estimates along with other changes aimed at increasing the new Volt's efficiency and performance.

**Comparison of the 1.4-L Gen-1 Volt engine on left with 2016 Gen-2 1.5-L on right. Note absence of fat orange 400-A cables on the new unit, due to integration of TPIM into drive unit.**

Nitz and the other technical leaders on hand—including Electric Drive Engineering Director Pete Savagian, Executive Chief Engineer Pam Fletcher, Chief Engineer Andrew Farah, Electrification General Director Tim Grewe, and GM's Battery Systems Director Bill Wallace—are veterans of the first-generation Volt program. They noted other benefits resulting from the 2016 improvements, including 20% stronger low-speed acceleration and quieter, smoother operation when in "range extender" mode with the gasoline engine engaged. These are a result of redesigned electric motors (two per drive unit) and clutch array, a new, higher torque 1.5-L four-cylinder combustion engine that replaces the previous 1.4-L unit, and a revamped TPIM (traction power inverter module).

The new Voltec system also is 130 lb (59 kg) lighter than the first-gen system, thanks in part to the ICE's aluminum cylinder block (the incumbent 1.4-L block is iron), the direct mounting of the TPIM to the transmission case which eliminates the heavy orange 400-A cables, and an overall focus on systems integration. Farah, who was disappointed in the original Volt's 3781-lb (1715-kg) curb weight, indicated that the 2016 car will be lighter overall as a result of these and other measures.

**The Gen-2 Volt battery pack is a clean-sheet design. It retains the previous pack's 300-V rating. It will have greater than the Gen-1's 16 kW-h energy storage capability and will be able to utilize more state of charge than the 65% available in the Gen-1 pack, engineers said.**

## New Voltec system

Greater range (particularly in EV mode), fuel efficiency, and power were the top three requests from owners of 2011-2013 Volt models, Nitz said. He explained that 60% of Volt buyers sign on to a GM program that funnels customer-use data directly to GM Engineering—"our secret weapon," Nitz chuckled. Volt owners do more than 80 percent of their driving in EV mode, and his team was surprised to find that owners charge their cars' batteries on average more than once per day, rather than the expected overnight charge. This indicates greater access to curbside charging at their destinations and a general desire for more battery energy-storage capacity.

**GM's top transmission engineer, the enthusiastic Larry Nitz, played a key role in development of the original 2011 Volt. He's leading the Gen-2 program.**

**Executive Chief Engineer Pam Fletcher said the Gen-2 Volt's onboard charging system will be lighter weight and be capable of delivering a full state of charge in 4 hours at 220-V.**

New electric pumps for the Gen-2 Voltec drive on a workbench at GM's Warren transmission plant.

Close-up of GM-designed Hitachi electric machine for the Gen-2 Voltec showing rectangular-bar windings.

Details of the Gen-2 Voltec transaxle drive mechanism.

Chief Engineer Andrew Farah will present the 2016 Volt vehicle at the 2015 Detroit auto show. He's rarely seen with such a staid expression.

GM General Director of Electrification Tim Grewe said his team looked at 1.0-L and 1.4-L turbo engines before setting on the naturally-aspirated 1.5-L for Gen-2 Volt use.

"They want greater EV operating capability" than the typical 35 to 40 miles per charge in the first-generation car, Nitz noted.

Many observers of Volt's development, including the author, expected GM to use its recently introduced 1.0-L global Ecotec three-cylinder gas engine as the second-gen Volt's range-extender generator. But a triple "presents some NVH challenges, particularly at start-up, in an electrified vehicle where noise and vibration are not welcome in the drive experience," explained Nitz. While the team's strategy with the original car was 'full-size battery with a half-size engine', sticking with naturally-aspirated four-cylinders and adding 100 cc of displacement provide the optimum balance of refinement and the additional torque customers crave, at moderate cost.

GM claims the noise intensity of its new 1.5 L four is up to 50% quieter than Volkswagen's EA211 1.4-L four and up to 25% quieter than Ford's 1.0-L turbo three-cylinder. The Ecotec features 12.5:1 compression, cooled EGR, and a new variable-displacement oil pump. It is certified to run on 87 octane regular-grade gasoline, according to Grewe.

In creating the Gen-2 Voltec drive system, engineers achieved a 60% volume reduction in the power electronics and inverter, not counting the rest of the hardware, noted Savagian, compared with the existing 4ET50 electric transaxle, along with a 5-12% efficiency gain overall and a 2% increase in motor efficiency when the electric machines are operating as generators. There was also a significant focus in reducing the use of rare-earth materials.

Pointing to a sectioned TPIM on display, Savagian noted the electronics use double-sided cooling on a custom automotive-grade package for the IGBTs (insulated gate bipolar transistors). "In past technologies we used something more akin to industrial-type IGBT packages," he explained. "And making the unit smaller means it's also more rigid which is good because the transmission-mounted environment is more severe—up to 6 times greater vibration frequencies than previously."

The twin electric machines, designed by GM and manufactured at Hitachi's Kentucky facility, use the bar-wound rectangular wire technology and tooling common to GM's other e-motors developed in-house. The motors are oil cooled, and together they use 10% less steel than is used in a single motor o SAE Technical Papers for publication at the 2015 SAE World Congress; highlights will be presented at the 2015 SAE Hybrid & Electric Vehicle Symposium, February 10-12 in Los Angeles; go to http://www.sae.org/events/hybridev/. ■

*Lindsay Brooke*

# Next-gen NSX: a twin-turbo, multi-material Ferrari-fighter

**The 2016 production NSX made its much-awaited global debut at NAIAS in January. More technical details of Acura's mid-engine AWD hybrid supercar will be shared by Honda R&D experts at the SAE 2015 World Congress.**

Three years removed from its concept-design debut, the Acura NSX supercar took to the 2015 North American International Auto Show (NAIAS) stage once more, this time in its production form. While recognizable as that concept's progeny, the next-generation 2016 NSX revealed this January in Detroit has evolved over its 36-month development program.

The Ohio-based, American-led R&D team for the new NSX, with Ted Klaus, Chief Engineer and Global Development Leader, at the helm, has made design and engineering decisions both big and small—but none insignificant—every step of the way.

Perhaps the biggest of these decisions, inevitably making exterior design changes necessary as well, was to abandon the initially planned naturally aspirated V6 engine for a power plant with more "growl." Which is something the new $150,000+ supercar will need plenty of to compete against what Klaus considers its key competitors: the Porsche 911 Turbo, Audi R8 V10, and "looking up toward" the Ferrari 458 Italia.

## Revising the powertrain on the fly

"Some of you may recall [the] blue NSX Concept that lapped Mid-Ohio before the start of the Indy Car race in August of 2013. This prototype was powered by a transverse-mounted, normally aspirated V6. It was a good powertrain, but for NSX good is not good enough," Klaus explained after the production NSX rolled onto the Detroit stage.

"So, in the middle of development, we made a bold decision to create an all-new longitudinally mounted, twin-turbocharged V6. Changing the powertrain design and layout was not an easy task—it was like undergoing a heart transplant while running a marathon. But 10 seconds behind the wheel, and you will understand how this new power unit is one of the keys to delivering a 'new sports experience.'"

The mid-mounted, 75° DOHC V6 engine (Klaus would not reveal the displacement) employs a "race-inspired" compact valvetrain and dry-sump lubrication system to help lower the center of gravity. It is paired with a new nine-speed dual clutch transmission (DCT) "of our own design," stated Klaus. The unit houses an electric motor that helps directly drive the rear wheels, to support acceleration, braking, and transmission shifting

**Outward visibility and straightforward interfaces were priorities for the interior design team.**

performance. Twin electric motors in front provide instantaneous torque delivery and dynamic torque vectoring. The Sport Hybrid system is said to generate "north of 550 hp."

"It will run on electric alone for short periods of time," said Klaus. "It is not a heavy plug-in hybrid…We have just the right amount of electric capacity to support the driver when they're demanding it but also to quickly recover energy from the brakes. It's a very lightweight, high-performance [system]—much closer to an F1 type."

> Honda, as host company of the SAE 2015 World Congress, is taking advantage of the venue as the next step in the launch of its 2016 Acura NSX supercar. Repeatedly during an interview with *Automotive Engineering* immediately following the production NSX's Detroit debut in January, Chief Engineer Ted Klaus referenced the SAE engineering event, taking place April 21-23 in Detroit, as the place where more technical details would be revealed.
>
> Asked about the use of advanced materials and where specifically they're employed in the car, Klaus said, "We're going to talk at length certainly as we move toward the SAE event, but each material is used strategically for its correct location."
>
> And again regarding the advanced manufacturing methods developed for the new NSX, he replied, "This space frame combines all of these materials as well as some world's-first processes that again we're going to discuss at length at this year's SAE [Congress]. And we'll certainly then have technical papers out there to help the global media to understand the details as we [move toward] media drives later this summer."
>
> Honda R&D plans to have several technical papers related to NSX technology as part of the SAE technical sessions, and the vehicle itself will be on display in the Cobo atrium, along with other Honda innovations.
>
> In addition, SAE members and guests can hear presentations from key members of the Honda R&D team involved with NSX development on April 22 at the SAE Detroit Section meeting, taking place at Cobo Center.
>
> *Ryan Gehm*

While vehicle development was led by Honda R&D Americas, engineering of the sport-hybrid power unit was executed in Japan, Klaus shared. "We were interdependent; we're a global team," he said. "The engine will be built in America at Anna [Ohio]. It's all about leveraging our global capabilities—certainly in Japan, in Europe just refining the product, and in America, we've integrated everything. And also working with our global suppliers and bringing on new suppliers." (Supplier information will be released at a later date, Klaus said.)

Key powertrain components including the Sport Hybrid battery pack and power control unit were optimally placed to concentrate vehicle mass low and toward the center of the vehicle.

"The combination of its lightweight chassis and low-mounted Sport Hybrid power unit, give the NSX the lowest center of gravity in its class," Klaus claims.

The NSX features a fully independent, all-aluminum front and rear suspension, riding on Continental ContiSportContact high-performance tires—245/35Z R19 front and 295/30Z R20 rear—with 19- x 8.5-in front and 20- x 11-in rear aluminum alloy wheels. Six-piston front and four-piston rear monoblock calipers and carbon-ceramic brake discs bring the supercar to a stop.

Sport Hybrid Super-Handling All Wheel Drive (Sport Hybrid SH-AWD) enhances stability, control, and launch performance while providing quicker response to driver inputs for steering, braking, and throttle. Agile Handling Assist (AHA) subtly applies brake torque to further enhance yaw response and dynamic stability.

An all-new longitudinally mounted, twin-turbocharged V6 engine is mated to a nine-speed DCT. The unit houses an electric motor that helps directly drive the rear wheels; twin electric motors in front provide instantaneous torque delivery and dynamic torque vectoring.

"The first NSX meant this, and the new NSX means this—the product to unleash the passion, the talents of our entire company. So inside of this [car] you have dreams realized in a very concrete way," said the man who led global development of the supercar, Ted Klaus.

The NSX's dial-operated Integrated Dynamic System (IDS) features four modes—Quiet, Sport, Sport+, and Track—plus a "launch" function. The system adjusts engine, motor, transmission, and chassis response accordingly, as well as the engine sound level. Quiet mode enables electric-only driving at lower speeds. Dynamic vehicle responses become "increasingly sharp" moving from Sport to Sport+ to Track.

"These guys are real time developing the car," Michelle Christensen, NSX Exterior Design Project Leader, said at NAIAS in January. "We're still taking it on the track, but can get rid of that camo a little bit now."

## Aerodynamic sans active elements

To accommodate the new longitudinally mounted twin-turbo V6 and nine-speed DCT, the production NSX is longer—by 3.2 in (80 mm)—and wider—by 1 in (25 mm)—than the NSX Concept shown in 2012. (See table for dimensions comparison between new NSX, the concept, and 2005 NSX.)

A slightly more cab-forward package also stems from the powertrain change, according to Christensen, who is based in the Torrance, CA, Acura design studio.

"Closely. Daily," Christensen said when asked how closely she worked with engineering on the exterior design of the vehicle. "I'm in the Torrance office and our engineers are in Ohio, but [we had] daily communication and I would fly out to Ohio for a week at a time."

The exterior body design was sculpted with "total airflow management" in mind, for optimum downforce and vehicle systems cooling—all with no active aerodynamic elements "because we wanted it to be pure and not add any extra gear and weight that we didn't really need," Christensen said. She gave *Automotive Engineering* a vehicle walk-around on the Detroit show floor, pointing out the many aero features and design changes.

"Compared to the 2012 show car, the biggest difference on the exterior is all the cooling and airflow elements. So that's really the pivotal styling change for us," she said. "When we changed from the transverse engine that we had earlier to the longitudinal twin-turbo, obviously now we need a lot more air intake. And as much air that's going in we need that to go out, so we took it to the wind tunnel [in Raymond, OH] and started exploring."

The NSX has three radiators in the front. Explaining the modifications made to the hood vents from concept to final body design, Christensen said, "They were further back on the original show car, and by doubling the size and pulling them forward, we were able to strategically get the air to land on the windshield and provide downforce; downforce is obviously critical for this car."

Other significant changes to the production car include new front fender vents, modified side air intakes, and an optimized deck spoiler.

"In the wind tunnel, we have a string test—it's kind of like the paint blot test, where you can see the direction of the airflow," Christensen said, discussing the efficacy of the fender vents. "We stuck the string

**2016 NSX development team leaders (l to r): Michelle Christensen, exterior design; John Norman, interior design; and Clement DeSouza, manufacturing.**

in the wheel well just to make sure it was in fact exiting out that side vent. As we put the string in there, it kind of wiggled its way out of that vent, stuck down the body side, and shot right into the side intake, which is awesome—that's right where we need all the air to go at high velocity."

The side intake and floating C-pillar collect air to feed the intercoolers and mid-mounted engine, and direct airflow over the rear deck to increase downforce.

"As the air comes out onto the decklid, it meets up with the air coming off the roof," she continued at the rear of the car. "So we were getting a lot of lift back here. We extended the spoiler a great deal from the original show car, and we also adjusted all these surfaces on the decklid to help guide the air; we had slightly different surfacing before." A small vent was also added to duct air out of the wheel well to reduce lift.

"This whole relationship is creating a lot of downforce, with the diffuser and the air coming out of the rear. It's all a very delicate balance," Christensen said. "All the updates that we did in the wind tunnel, I think it changed the styling for the better, made it a little more aggressive, more exotic."

## An interior that 'disappears'

While the exterior surely will attract attention, the interior of the new NSX—with its "human support cockpit" design theme—is meant to not be noticed. "We didn't want to create a glitzy, elaborate interior. We wanted to make sure that the interior almost disappeared when you drove," John Norman, NSX Interior Design Project Leader, shared with *Automotive Engineering*.

"I had this excellent experience when we did the original ride-and-drive for the car during the development," he continued. "We had all of our competitors; we had a Lamborghini there, a Ferrari, Porsche, the Nissan GT-R, a [Chevrolet] Corvette, we had all these fantastic cars. But we also had an original NSX to drive. Now while the performance benchmark had gotten much more modern and much faster, I read about driving the original NSX, how it was like the invisible car, and I actually experienced that. We were driving through the canyon roads of Malibu and I had this transcendent driving experience where everything just went away, the car went away, and it was just me and the road. And I said, 'Wow! We really need to keep this [with the next-gen NSX].'"

To that end, designers focused on providing great forward visibility; simple, intuitive controls; and "class-leading" ergonomics—most notably with the seats, which feature "top-class" holding performance thanks to suede inserts and offer easy ingress/egress.

"We really focused on pushing the IP [instrument panel] down as low as we could, making the pillars as thin as possible, prioritizing outward visibility. Even though it's a very low, wide, very sexy car, when you sit in it, it doesn't feel completely like you're inside the space shuttle or some really compromised seating position."

The NSX's instrument cluster features a dynamic thin-film-transistor display that responds to changes in the driver-selectable IDS with pertinent graphics and information. The Power button resides in the center of the new IDS dial control.

"The interface is relatively straightforward," Norman said. "In some of our other Acura cars we have the high screen, but from a package standpoint we wanted to really prioritize outward visibility, so we have a low-mounted touchscreen. Also on our current cars, the IDS control function is just a small button behind the shifter that you cycle through the modes; this one, we turned it into this large rotary knob and placed it kind of high in the center panel…We made it big enough that it can function at track speed—you can just grab it with a gloved hand and select your mode."

**The NSX features "world-first" casting technology for aluminum frame components that combines the design and manufacturing flexibility of a casting process with the strength and elongation properties of a forged material.**

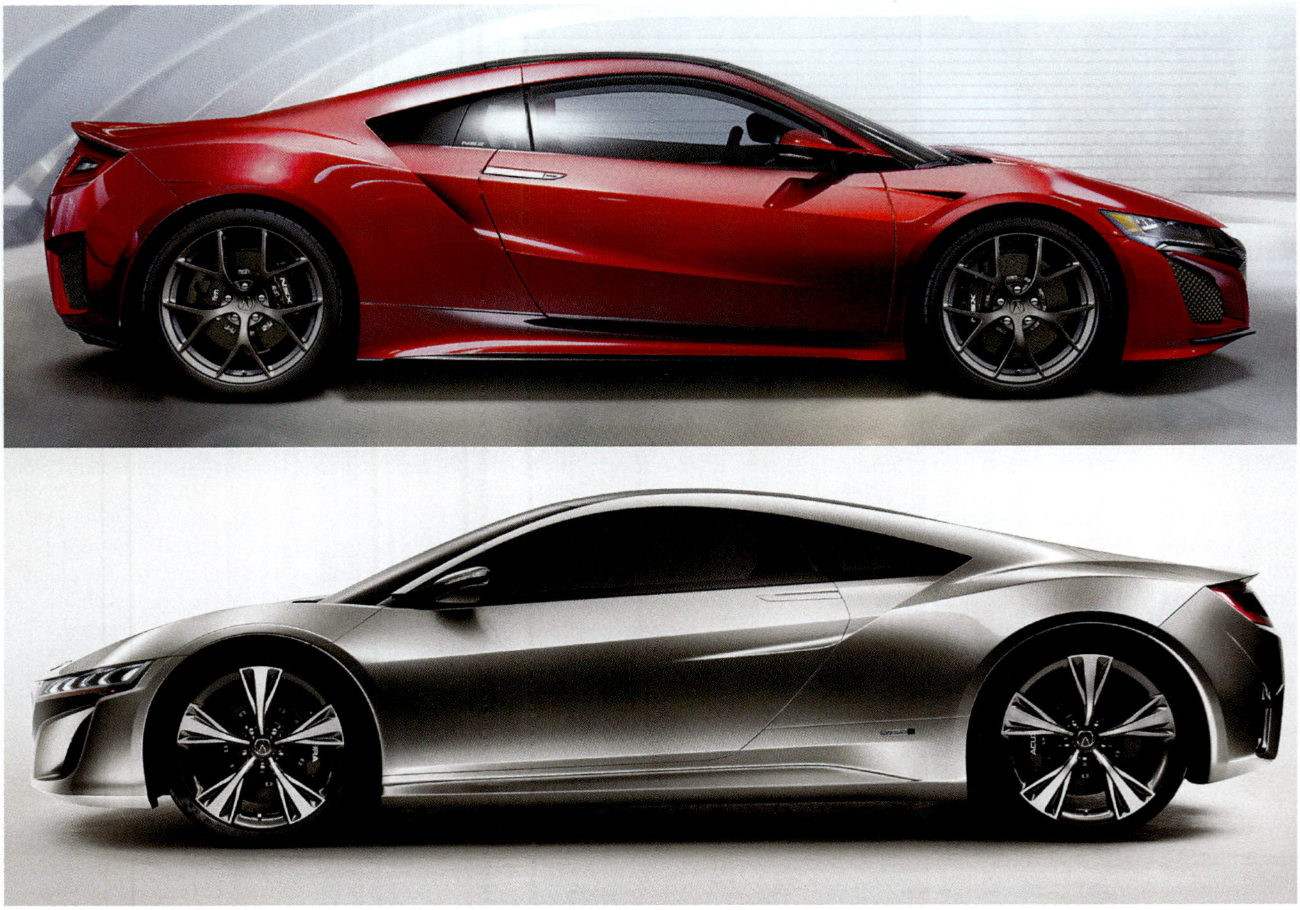

**To accommodate the new powertrain layout, the production NSX (top) is longer by 3.2 in (80 mm) and wider by 1 in (25 mm) compared to the NSX Concept (bottom). Designers also extended the spoiler from the original show car to aid downforce.**

Below the handcrafted leather dash panel is an exposed mid-frame—a functioning chassis structural member—made of a structural polymer with metal plating, according to Norman. "We didn't want to put some giant metal casting in there; that'd just be ridiculous [in terms of weight]," he said, noting that parts consolidation and weight reduction were major goals of the interior development program.

## Manufacturing—a materials technology-enabler

After stating that there "absolutely" will be a right-hand-drive version of the NSX since it's a global project, Klaus gave a "shout-out" to the new Performance Manufacturing Center (PMC) in Marysville, OH, where the supercar will be exclusively manufactured, as well as to the man leading the creation of the plant, Clement DeSouza.

"We would not have been able to deliver this product without this new factory," Klaus said. "To achieve that styling, to achieve our performance, the story of the factory enabling new technologies is a very exciting [one]."

The multi-material chassis comprises an aluminum-intensive space frame complemented by the "strategic use" of ultra-high-strength steel (UHSS) and carbon fiber. UHSS is used for the thin A-pillar, which is a continuous member, while carbon fiber is placed at the central floor, "close to the spine of the car," Klaus said, to enhance torsional and bending stiffness.

"I know ultra-high-strength steel maybe isn't as sexy as carbon fiber or aluminum, but if it's the most effective way to meet your overall design concept, that's what you should be doing," said Klaus. "[Plus] it's lighter than if we tried to use aluminum [for the A-pillar] and it's not as strong but way more robust than using carbon fiber."

Body panels are composed of aluminum and sheet molding composite.

The NSX features "world-first" casting technology that combines the design and manufacturing flexibility of a casting process with the strength and elongation properties of a forged material, enabling significant weight reduction, according to DeSouza.

"One of the big reasons we picked Ohio [for manufacturing] was because of proximity to R&D and our ability to work with them on a daily basis to realize this model," he told *Automotive Engineering*. Development of the casting technology provides "one of our biggest material differences and also biggest challenges on this project."

SAE Eye on Engineering Quick Take:

## Keeping NSK cool

Watch the video at: http://youtu.be/9zEuWslVwOU

Engines and transmissions create heat, and that challenges engineers to manage it so the parts inside don't fail. In this episode of SAE Eye on Engineering, Senior Editor Lindsay Brooke looks at Acura's 2016 NSX and its cooling system. SAE Eye on Engineering can be viewed at http://youtu.be/9zEuWslVwOU. It also airs in audio-only form Monday mornings on WJR 760 AM Detroit's Paul W. Smith Show. Access archived episodes of SAE Eye on Engineering at www.sae.org/magazines/podcasts.

Access archived episodes of SAE Eye on Engineering at www.sae.org/magazines/podcasts.

SAE Eye on Engineering Quick Take:

## Honda's U.S. engineering

Watch the video at: https://youtu.be/fwarDAh5IGU

More and more, Honda vehicles are getting their start on the CAD screens of American designers based in central Ohio, rather than those in Japan. In this episode of SAE Eye on Engineering, Senior Editor Lindsay Brooke looks at the busy year for Honda engineers in Raymond, OH. SAE Eye on Engineering can be viewed at https://youtu.be/fwarDAh5IGU. It also airs in audio-only form Monday mornings on WJR 760 AM Detroit's Paul W. Smith Show. Access archived episodes of SAE Eye on Engineering at www.sae.org/magazines/podcasts.

Access archived episodes of SAE Eye on Engineering at www.sae.org/magazines/podcasts.

The casting is employed for aluminum frame parts, but the process will be carried out at the Anna engine plant. "We had to leverage our casting experience at Anna engine, so we decided to do it there," DeSouza shared.

SAE technical paper 2015-01-0512, "Cast body nodes for 2016 NSX," will be presented as part of the "Advances in Lightweight Materials" session on April 22 at the SAE World Congress in Detroit (see sidebar). The paper details ablation casting, described as "an emerging technology which combines traditional sand molding techniques with rapid cooling due to the use of a water soluble binder." Six different body node castings have been manufactured for the aluminum space frame using ablation. These castings are integrated into the crash structure.

**Acura NSX - Dimensions Comparison**

| | New 2016 NSX | 2013 NSX Concept | Difference | 2005 NSX | Difference |
|---|---|---|---|---|---|
| Length, in (mm) | 176 (4470) | 172.8 (4390) | +3.2 (80) | 174.2 (4425) | +1.8 (45) |
| Width, in (mm) | 76.4 (1940) | 75.4 (1915) | +1.0 (25) | 71.3 (1810) | +5.1 (130) |
| Height, in (mm) | 47.8 (1215) | 47.2 (1200) | +0.6 (15) | 46.1 (1170) | +1.7 (45) |
| Wheelbase, in (mm) | 103.5 (2630) | 102.8 (2610) | +0.7 (20) | 99.6 (2530) | +3.9 (100) |
| Front track, in (mm) | 65.2 (1655) | --- | --- | 59.5 (1510) | +5.7 (145) |
| Rear track, in (mm) | 63.6 (1615) | --- | --- | 60.7 (1540) | +2.9 (75) |

Full body construction, paint, and final assembly will be conducted at PMC, where about 100 associates will help support and perform those tasks. And DeSouza notes that it's not your typical small-volume operation.

"It's a blend of craftsmanship and technology," he explained. "Usually when you go to a small-volume facility, you see a lot of hands-on work, but… keeping quality in mind and the demands of this vehicle, we realized we could use technology. Looking at each process and determining, 'Do we have a robot do it or have a person do it?'—that was our big driver in figuring out what we would devise."

The body shop is one area that benefits from the deployment of technology. "If you look at most of the other supercars, a lot of them are hand-MIG welded, but we're going to 100% robotic welding for the body," DeSouza said. "That helps guarantee the quality, the heat distribution, the accuracy of the body, and torsional rigidity. [With] aluminum, the biggest challenge is heat distortion...It gives us so much more flexibility by using a robot."

The company will begin accepting custom orders for the new NSX starting this summer. "There will be more than one spec because we know this customer wants to be able to configure, customize, and build their own vehicle," said Mike Accavitti, Senior Vice President and General Manager, Acura Division.

Klaus acknowledges that the supply-and-demand equation may cause some disappointment. "We are conscious that allocating the products globally is going to be a challenge," he said. "Especially if we choose to not satisfy demand…We have not given production figures and will not, but there is only so much we can do."

Production is expected to commence at PMC in the fall, with first deliveries planned for later in the year. Klaus and his team are quite certain those customers driving away in the new NSX will not be disappointed. ■

*Ryan Gehm*

# 2016 Explorer gains more powerful four-cylinder, Platinum opulence

Fresh front-end sheet metal, LED headlights, and active grille shutters are highlights of the 2016 Ford Explorer.

Ford has relentlessly applied all of the technology at its disposal in a program of continuous refinement to the 2016 Explorer in a bid to keep competitive a vehicle whose platform roots can be traced back through the 2005 Freestyle to its origin in the 1998 Volvo S80 sedan.

The primary advance for 2016 is the replacement of the 2.0-L EcoBoost four-cylinder as the fuel-saving upgrade engine for base model, XLT, and Limited-grade Explorers, by a freshly upgraded version of the 2.3-L EcoBoost seen previously in the Mustang and Lincoln MKC. The base engine remains a 290-hp (216-kW), 255-lb·ft (346-N·m) naturally aspirated 3.5-L V6.

According to Chief Program Engineer Arie Groeneveld, the Explorer's version of the new 2.3-L engine enjoys a variable oil pump that works only hard enough to produce the needed oil pressure, saving wasted effort and the corresponding fuel. It has a higher compression ratio, an active wastegate on the turbocharger, and polished valve buckets for reduced friction.

This engine will be rated at something more than 270 hp (201 kW) and 300 lb·ft (408 N·m), though official numbers were not yet available. Critically, this added power in comparison to the previous 2.0-L base engine means that Ford can offer all-wheel drive with the four-cylinder engine for the first time, and it can certify the Explorer with a 3000-lb (1360-kg) towing rating with its base engine.

This will be critical to sales, as many SUV buyers want the ability to drive in inclement weather, do light off-roading, or periodically tow, noted Groeneveld. Previously, 65% of Explorers were equipped with all-wheel drive, but that number will likely climb now that it is available with the most efficient engine, he said.

Despite the increases in displacement, power, and capability, the 2016 Explorer should match the EPA fuel economy ratings of the 2.0-L 2015, which topped out at 28 mpg EPA highway, he added.

In keeping with the theme of upgrading the model line, in addition to an improved optional engine for the popular trim levels, Ford is pushing the top of the Explorer's line upward, with the creation of a Platinum edition that is packed with still more technology and comforts.

For the outgoing model, 90% of shoppers who bought the Sport trim level loaded it with all the options. "That's a strong signal customers are ready for Platinum," pointed out Matt Zuehlk, Explorer Brand Manager.

The 10-inch display embedded between the analog gauges is a first for Ford.

The Explorer's cabin's Lear seats are swaddled in Nirvana leather, the softest ever fitted to a Ford brand product, according to Groeneveld. Occupants will be bathed in the sound from a Ford-first Sony 500-watt audio system with Live Acoustics and Clear Phase technology. Live Acoustics is said to recreate the sound of a concert venue, while Clear Phase eliminates sound dispersion in the vehicle.

And Ford's first use of a 10-in instrument cluster display flanked by analog gauges greets drivers when the slip behind the wheel.

They will also enjoy the benefit of 180-degree cameras front and rear for easy parking, and those are equipped with industry-first lens washers. Customers in Snowbelt areas and rural drivers will appreciate not having to constantly wipe lenses clear, Groeneveld promised.

Another contributor to the Explorer's efficiency is the installation of active grille shutters for reduced aero drag at highway speeds. The 2016 wears new sheet metal everywhere except the roof and doors, with crisp lines lending a contemporary flair.

The strong horizontal theme in the front end emphasizes width and minimizes the front overhang, which is increased slightly to accommodate the grille shutters. The headlights have standard LED low beams and LED running lights and are housed in sleek rectangular reflectors that simultaneously resemble quad-lamp sealed beam headlights.

**The Explorer's Lear seats are wrapped in leather the company says is Ford's softest ever.**

Integration of so many new components demanded a lot of time, which represented the engineering team's biggest challenge when developing the 2016 Explorer, Groeneveld reported. "To get the attribute level we were looking for takes a lot of time in the vehicle," he said. ■

*Dan Carney*

# Ford GT prototypes "run naked" in public-road testing phase

Dogging it in the right lane of I-696 west of Farmington, MI, a Ford GT prototype shows its aerodynamically sophisticated bodywork.

Without camouflage or dazzle paint jobs, Ford's 2017 GT development program has begun public-road testing in southeastern Michigan. The exotic American supercars are "running naked" as they rack up miles amid daily commuter traffic. *Automotive Engineering* encountered and photographed one of the off-tool prototypes recently, shown here moving at a sedate 70 mph (113 km/h) on I-696 westbound near Novi, MI.

Even in its flat-gray paint, the carbon-fiber and aluminum GT (wearing a manufacturer's license plate but no badges) had nearby drivers' heads spinning. The car is stunning when viewed from the rear quarter, its twin "flying buttress" airfoils linking the roof to the muscular, motorsport-influenced rear fenders. Besides being designed to create downforce, the buttresses also vector air toward intercoolers integrated into the leading edges of the fenders, which eject hot air toward the rear of the car, exiting through the taillamp centers.

As shown in the photos, the prototype was missing its production taillamps, replaced by Pep Boys-type brake lights. Also not fitted on the test car was the rear lower fascia, exposing the large, transverse-mounted exhaust muffler system. The multi-axis active rear spoiler, while not deployed, is designed to provide both downforce and drag.

"Development of the GT continues unabated," a Ford engineering source tells *Automotive Engineering*, "and early development mules are now rolling consistently around Dearborn in their bare carbon-fiber body panels in advance of the anticipated launch late next year."

Concurrent with development of the production GT—due to be priced around $400,000 with volume limited to 250 units/year—is the GT racecar program. Ford is working with long-time racecar partner, Canada-based Multimatic on both programs and is committed to running a pair of cars in each of two major

The 2016 GT represents Ford's most extensive use of CFD analysis on a production vehicle to date, said Ford Performance Chief Engineer Jamal Hameedi.

race series, the IMSA/TUDOR (U.S.) and WEC (World Endurance Championship). The latter includes the 2016 LeMans 24-hour epic, marking 50 years since the original GT40's Ferrari-beating win there.

The new GT racecar is being tested at Calabogie Motorsports Park near Multimatic's Ontario engineering facility, according to a Ford engineer familiar with GT development. He said Roush Yates Engines is leading development of the turbocharged V6 race powertrain. The racecar test program will "hit the tracks in late autumn" scheduled for Sebring, Daytona, and the Circuit of the Americas in Austin, TX.

For more SAE articles on the new GT, see http://articles.sae.org/13840/ and http://video.sae.org/12009/. ■

*Lindsay Brooke*

SAE Eye on Engineering Quick Take:

## Ford GT blockbuster at NAIAS

Watch the video at: http://youtu.be/s1gSJdoxyzQ

Every auto show needs a blockbuster vehicle intro to make the show great. And that's what Ford did last week, when it unveiled the new Ford GT supercar in Detroit. In this episode of SAE Eye on Engineering, Senior Editor Lindsay Brooke looks at Ford's unveiling of the new GT at the 2015 North American International Auto Show. SAE Eye on Engineering can be viewed at http://youtu.be/s1gSJdoxyzQ. It also airs in audio-only form Monday mornings on WJR 760 AM Detroit's Paul W. Smith Show.

Access archived episodes of SAE Eye on Engineering at www.sae.org/magazines/podcasts.

# Ford aims 2017 GT supercar at the street and Le Mans

**The 2017 GT, Ford's new carbon fiber supercar, will get the highest-output EcoBoost V6. It was designed to compete in the 2016 24 Hours of LeMans marking the 50th anniversary of Ford's first victory over Ferrari.**

Ford's ongoing development of a green image does not mean ignoring the performance market, and at the 2015 North American International Auto Show it introduced new editions of the Raptor pickup and Shelby Mustang GT350R, and dropped a genuine bombshell with the debut of the new Ford GT supercar.

First the head-spinning GT, which is a thoroughly modern interpretation of Ford's original GT40 factory race car that won the 24 Hours of Le Mans four consecutive years beginning in 1966. The new GT will enter production in street trim next year (as a 2017 model) as well as serve as Ford's platform for a Le Mans effort next year—marking the 50th anniversary of the GT40's first victory over Ferrari. Industry engineers familiar with the program tell *Automotive Engineering* that the new GT will be built by Markham, ON-based Multimatic Engineering, which has extensive experience in carbon fiber component manufacturing and enjoys a long relationship with Ford Motorsport.

Featuring a mid-engine/rear-drive layout and carbon-fiber monocoque chassis and 20-in wheels, the GT is powered by a 3.5-L twin-turbocharged Ecoboost V6 that Ford engineers promise will deliver more than 600 hp (447 kW) to a 7-speed dual-clutch automatic transmission. The engine benefits from learnings gained from Ford's racing V6 used in Daytona Prototype-class machines.

Remarkably, the show car represents approximately 12 months of development. Perhaps even more impressive in an era where nearly everyone carries a mobile camera with internet capability was Ford's ability to keep the car secret—not a single leaked web photo emerged—until its NAIAS unveiling on January 12. The show car's overall exterior form, with its flying buttresses that connect each rear fender to the car's roof and feature a downforce-generating airfoil geometry, is "about 95% finished," according to VP of Design Moray Callum.

The complex exterior is an exercise in careful computational work aimed at making the car aerodynamically efficient as well as enhancing engine performance—intercoolers integrated into the leading edges of the F1-influenced rear fenders eject hot air toward the rear of the car, exiting through the taillamp centers, explained Jamal Hameedi, Global Performance Vehicle Chief Engineer.

**Raptor is the F-150 halo vehicle that fits right into Ford performance lineup with its high-performance EcoBoost V6 replacing the previous V8.**

The design includes an active, multi-axis rear spoiler that provides both downforce and drag when it deploys to full vertical extension.

The carbon-fiber tub connects to aluminum subframes front and rear. Front suspension is by a pushrod-type arrangement. The show car wears carbon-ceramic brake rotors and Brembo calipers. The car's doors swing upward to open to optimize ingress and egress. Inside the cockpit the two seats are non-adjustable, being integrated into the composite tub. The pedal cluster and steering wheel are electrically adjustable, however, and the instrument cluster is reconfigurable.

The high-performance V6 features a dual fuel injection system with both port-facing and direct injectors. Ford engineers said dual injection permits use of smaller nozzle injectors in each of the two systems to more precisely control the fuel flow under all conditions. This contrasts with simply using injectors with larger holes, that would allow the increased fuel flow necessary but would lose precision, particularly when low fuel flow rates are commanded.

*Automotive Engineering* will provide further reports of the GT's development, technology, and race preparation in the months ahead.

## All-new aluminum Raptor

The Raptor may be based on the F-150, but the new second generation continues to be a full-fledged member of the Ford high-performance family and serves as the "halo" of the F-Series family, according to Doug Scott, Truck Marketing Manager. The new Raptor gets an aluminum body like its F-Series cousins and a new higher-strength steel frame. The vehicle is about 500 lb. (227 kg) lighter than the previous model, vs. 700 lb (318 kg) for the F-150, as Ford "gave back" some weight on the Raptor to provide additional feature content and upgrade the frame.

Like the GT, the new Raptor is powered by a 3.5-L EcoBoost V6 that makes more power than the previous model's SAE-rated 411 hp (307 kW) V8, according to Scott. A true dual exhaust is part of the new engine package. Although the displacement is unchanged, the twin-turbo V6 itself is almost all-new, Scott said, with only a handful of carryover parts. It will be phased in across the board. This is also the first application of the new 10-speed planetary automatic that Ford is developing jointly with GM. The transmission is mated to a new two-speed transfer case with a novel integration of its operation.

**Track-ready Shelby Mustang GT 350 R remains a V8, and is built without such accoutrements as A/C, stereo, rear seat, etc. to save weight.**

Unlike the previous model with just 2WD, 4L and 4H, the new Raptor adds an AWD that is based on variable clutch operation. However, when the system automatically goes into 4L or 4H, like the driver-selectable gearing it mechanically locks into position. The 2WD mode is driver-selectable or is very close when the system senses an appropriate load condition. The case is a BorgWarner design, exclusively for the Raptor at this time.

The Raptor gets a terrain management system with six modes: sand, on-road, sport, normal, rock crawl and "baja" (as in high-speed desert operation). Unlike others that are selectable by a turn knob, the Raptor's appears on the cluster for driver to make a pick.

The vehicle is 6.0-in (152 mm) wider than the F-150, has more suspension travel, and will be available in a Super Cab and Super Crew. All the F-150 safety-convenience features, from adaptive cruise to cargo box lighting and 360° camera also are available on the Raptor.

The Fox high-performance off-road racing shocks continue, but the new Raptor specs a larger piston size: to 3.0-in (76 mm), from the previous 2.5-in (64 mm).

## Shelby Mustang GT 350 R keeps a V8

There is one new Ford performance variant, however, where the V8 remains king: the Shelby Mustang GT 350 R, a track-ready model that reflects a truly minimalist approach. Ford took out the rear seat, stereo, trunk floorboard and carpet, backup camera, exhaust resonators, even the tire sealer and inflator, and of course deleted the air conditioning. It is equipped with carbon fiber wheels, alone saving some 50 lb (23 kg) vs. aluminum, and the car is 130 lb (59 kg) lighter than the previous model.

A major upgrade is the new 5.2-L V8, which will be SAE-rated at over 500 hp (373 kW). It's mated to a 6-speed Tremec manual. The V8 features Ford's first flat-plane crankshaft (see http://articles.sae.org/13709/), which features crankpins spaced at 180° rather than 90° on a conventional cross-plane crank. Cylinder firing alternates from one bank to the other, which reduces overlap of exhaust gas pulses. In conjunction with changes to the valvetrain calibration and cylinder head, the result is improved high-end engine breathing.

The GT350 R's body is designed for increased downforce, starting with a reshaped front fascia, plus a ducted belly pan and a subtle lip on the rear spoiler. Suspension is by continuously variable MagneRide (BWI Group's electronically-controlled damping with iron-powder-filled fluid), with a claimed 10-ms response time for high-performance operation. ■

*Paul Weissler and Lindsay Brooke*

SAE Eye on Engineering Quick Take:

### Ford GT blockbuster at NAIAS

Watch the video at: https://youtu.be/WCRRUhr_9w8

Sensors gather information about the vehicle, its performance, and the surrounding environment. In this episode of SAE Eye on Engineering, Senior Editor Lindsay Brooke looks at the sensors and microprocessors on the 2016 Ford GT. SAE Eye on Engineering can be viewed at https://youtu.be/WCRRUhr_9w8. It also airs in audio-only form Monday mornings on WJR 760 AM Detroit's Paul W. Smith Show.

Access archived episodes of SAE Eye on Engineering at www.sae.org/magazines/podcasts.

# Flat-plane crankshaft spins 'Voodoo' into Ford's new Shelby V8

**This is the view of the 2015 Shelby GT350 that Ford engineers predict most Camaro Z-28 and Porsche 911 GT3 owners will see when the new hi-po Mustang hits the streets and track days early next year. Ford will release engine visuals at a later date.**

So you think the Corvette Stingray and Jaguar F-Type play the wildest V8 soundtracks through their active-exhaust systems? Or you say nothing beats a Ferrari 458 for pure aural delight? Well, you haven't heard Ford's 2015 Shelby GT350 Mustang on rising throttle. From tip-in to WOT the new Shelby, powered by a unique naturally aspirated 5.2-L DOHC V8 that was coded "Voodoo" during development, plays a downright diabolical tune from its four exhaust outlets.

"When we would start up the first prototypes in the engineering garage, everybody's heads turned: 'What the heck is that?'" recalled Raj Nair, Ford's Group Vice President of Global Product Development, in an interview with *Automotive Engineering* prior to the 2014 L.A. Auto Show. "When we'd transit the cars between facilities in Dearborn you'd hear their echo out on Oakwood Boulevard. Anybody at our proving ground who heard these cars knew there was something very different under the hood."

There will be customers who will buy the latest Shelby Mustang for its 20% stiffer body structure (compared with the Mustang GT), its racetrack-focused chassis tuning, enormous Brembo brakes, responsive MagneRide damping, Recaro seats, and many other vehicle attributes. Others, however, will want the car for its howling exhaust note alone. The sound is created by the engine's special flat-plane crankshaft and its related breathing and ignition characteristics.

So named because its four connecting-rod journals are spaced 180° apart, the shaft appears "flat" in a single plane when viewed head-on. The arrangement generally does away with the large counterweights used in the typical 90° "cross plane" layout which helps reduce crankshaft mass and enables the engine to rev more freely. The new Ford spins to a heady 8000 rpm before fuel cut-off, according to Nair, and has another 250 rpm of safe headroom. This from mildly oversquare 94 x 92.7-mm bore and stroke dimensions.

Long the province of European high-performance V8s from Ferrari, Lotus, Cosworth (the legendary DFV Formula 1 unit) and others, flat-plane cranks were also used in the first Cadillac V8 in 1914 . The configuration gives an alternating right-left-right-left firing order across the cylinder banks, allowing more efficient gas flow and exhaust-gas scavenging along with improved throttle response and exotic exhaust acoustics.

Compression ratio of the port-injected, Coyote-based 5.2-L is 12:1 and max cylinder pressure is 97 bar (1406 psi), an engineer said. Final SAE power and torque ratings are not yet finalized, but the V8 will deliver at least 500 hp (373 kW) and more than 400 lb·ft (542 N·m), according to Jamal Hameedi, Chief Engineer, Ford Global Performance Vehicles.

Hear the V8 sing in this video courtesy of Ford (turn up your volume setting first): https://www.youtube.com/watch?v=WxKhiCh9r0k

**Global PD boss Raj Nair had faith that the Powertrain and GT350 teams would solve the NVH issues related to the prototype V8. "We all wanted this motor," he said.**

Flat-plane-crank engines have limitations. First, the lack of counterweight balancing typically limits cylinder displacement to about 4.5-4.6-L due to greater second-order vibration. Ford has solved that in the 5.2-L application with a new crankshaft-mounted damper system and extraordinary attention to NVH abatement during the design and prototype phases. According to Nair, the engine program (which was concurrent with GT350 vehicle development) nearly wasn't approved for production.

"This [vibration] was our biggest engineering challenge even after we had the first prototype," Nair noted. "Things were breaking and the technical guys were worried. Whether or not we continued down the flat-plane-crank path for GT350 came down to a critical prototype drive we had scheduled. After that drive, we all went into a meeting room for the debriefing. And we unanimously concluded that we simply had to have this motor! We were determined to solve the issues."

What followed was, in Nair's words, "a lot of stiffening of the cylinder block, exhaust system, and various brackets" achieved through an intense FEA analysis, plus "a lot of tuning." Nair said the result yielded unique torsional-damping technology among other patent-pending actions aimed at taming the beast.

The 5.2-L is naturally aspirated by design; boosting was not part of the plan. "We really wanted to stay with naturally-aspirated technologies for instantaneous throttle response" along with a feeling of minimal flywheel effect, Nair explained. He's familiar with both qualities as a longtime motorcyclist who rides a Yamaha R1 sportbike on the street and at track days.

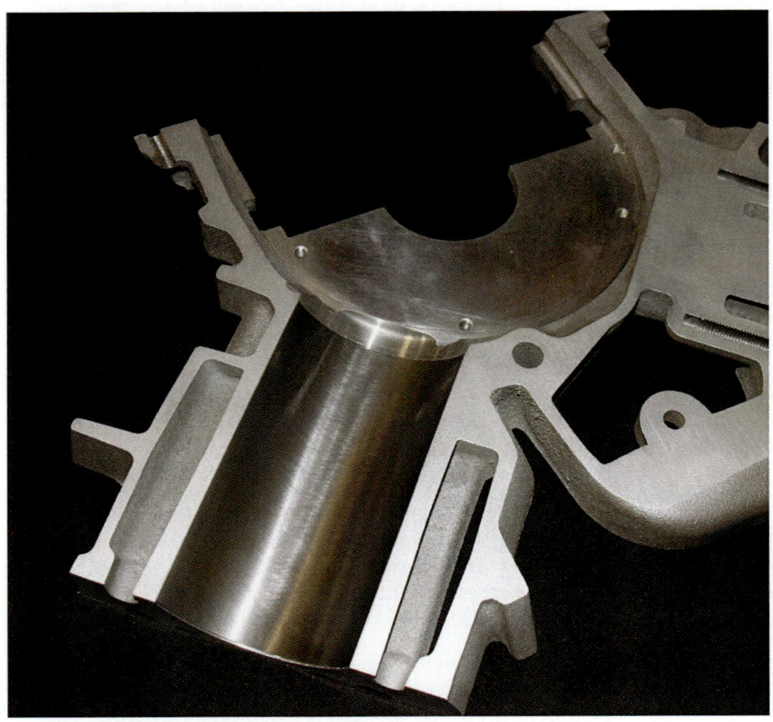

**Ford will use its patented PTWA cylinder bore coating technology (shown here on a cutaway of the 5.4-L Shelby V8) on the new 5.2-L engine. (Lindsay Brooke photo)**

Other challenges with flat-plane-crank engines include their tendency to deliver underwhelming torque. Nair admitted that the GT350 "doesn't have that electric-motor feel off idle that the GT500 has" but the latter also uses a blown 5.8-L rated at 631 lb·ft/855 N·m. Increasing cylinder displacement (via larger bore) to 5.2-L, which is 700-cc larger than Ferrari's V8 used in the 458, helped overcome some of the torque deficit. Nair also noted that optimizing the intake and exhaust systems for power, efficiency, and meeting cold-start emissions was tricky.

"Even the unequal-length headers we selected, to some extent, to help with catalyst light-off," he said.

According to Hameedi, the 5.2-L V8 has a linerless aluminum cylinder block that features bores finished using PTWA (Plasma Transfer Wire Arc), an advanced coating technology co-developed by Ford and Flame-Spray Industries that debuted on the 2011 GT500 5.4-L V8. For PTWA details see http://articles.sae.org/7624/.

Supplier sources involved with the 5.2L V8 program said Ford broadcast an initial volume plan for production of 5000 units.

*Lindsay Brooke*

# 2016 Civic targets Audi A3, drives new modular-engineering strategy

Beyond its sporty concept skin the 10th-generation Civic is in the vanguard of Honda's new approach to increased use of common component sets from an expanded list of suppliers.

When it launches this summer, Honda's next-generation Civic will debut a lot more than a sporty new design language and more efficient and spirited powertrains. The 2016 Civic serves as the lead application for Honda's new Global Compact platform that will also underpin the next-generation Accord as well as the HR-V crossover.

"It's a super-significant global engineering program," said an American Honda Motor Co. source who spoke with *Automotive Engineering* on the condition of anonymity. "Consolidating our two biggest vehicle platforms will dramatically boost production scale, by millions of units, across multiple nameplates."

Equally important will be a bold new modular component-set approach, most visibly in the structure and subsystems located ahead of the A-pillar on the new-generation vehicles, which will leverage a broader use of global suppliers.

"We're going way beyond our traditional keiretsu supply chain beginning with the '16 Civic," the source explained. "You'll see supplier names who haven't previously been with Honda on this program."

With this development Honda joins a growing list of competitors that have recently introduced similar common-component-set strategies aimed at balancing cost, production efficiency, rapid technology adoption, and quality.

The lime-green Civic Coupe concept unveiled at the 2015 New York International Auto Show (http://articles.sae.org/14025/) provided a hint of the features, technologies, and the compact-vehicle assault that Honda engineers are preparing to unleash in the coming months. While the zesty concept at the show did not have an interior, Honda designers and engineers aimed well beyond the Civic's typical competitive set for benchmarking interior materials and finish, cabin NVH, and occupant comfort, in addition to vehicle dynamics.

"Audi A3 was our primary bogey," the Honda source noted. "And I can say honestly that we've topped it with the 2016 Civic."

A Honda R&D veteran described the general atmosphere at the company's Raymond, OH, development complex as "electric and intense" as of early April.

"For the first time our North American engineering team is executing both the next global Civic and Acura NSX programs, virtually simultaneously," the R&D source said. "Lots of late hours and stress there, for sure. But the products are transformative." ■

*Lindsay Brooke*

SAE Eye on Engineering Quick Take:

## Honda's U.S. engineering

Watch the video at: https://youtu.be/fwarDAh5IGU

More and more, Honda vehicles are getting their start on the CAD screens of American designers based in central Ohio, rather than those in Japan. In this episode of SAE Eye on Engineering, Senior Editor Lindsay Brooke looks at the busy year for Honda engineers in Raymond, OH. SAE Eye on Engineering can be viewed at https://youtu.be/fwarDAh5IGU. It also airs in audio-only form Monday mornings on WJR 760 AM Detroit's Paul W. Smith Show.

Access archived episodes of SAE Eye on Engineering at www.sae.org/magazines/podcasts.

# Honda surprises with Civic Concept

**The Honda Civic Concept's attractive styling runs deeper than its retina-scorching green show car paint job, with a longer hood and shorter front overhang lending a sportier appearance.**

Honda's Civic is a critically important model for the company, one that has sold 20 million units worldwide and 10 million in the U.S. since its 1972 introduction. Nine million of those have been assembled in North America.

So when the current-edition Civic stumbled out of the gate in 2011, earning bushels of criticism from reviewers, the company resolved not to make the same mistakes on its successor, which debuted in concept form at the 2015 New York International Auto Show.

The outgoing car was seen as visually unappealing, with dreary dynamics and cheap appointments. Some of these shortcomings were shored up in a hasty refresh, but the car was fundamentally inferior to quality competitors.

Not this time, swears Honda. "In every way, this will be an epic Civic," promised John Mendel, Executive Vice President of American Honda. "The new Civic will be hands-down the most dynamic, the most technologically advanced, and the most refined and stylish Civic we've ever brought to market."

Although Honda only revealed the concept version of the all-new tenth-generation Civic in New York, the company did provide some detail of the car, which will first arrive in four-door sedan form in the fall of this year, to be followed by two-door coupes and five-door hatchbacks.

The 2016 Civic is built on a new platform that the company promises is more rigid and lighter than the current car. As with all of Honda's current North American models, the new Civic will be expected to ace tough crash tests from the National Highway Traffic Safety Administration and the Insurance Institute for Highway Safety, including the IIHS's brutal small overlap frontal crash test.

(Read more about the 2016 Civic's development program and new platform athttp://articles.sae.org/14040/.)

Honda will continue its effort to democratize safety by offering what it calls the "Honda Sensing" suite of active technologies on the Civic, to include lane-departure warning, forward-collision warning, rear-view camera, adaptive cruise control, and the company's LaneWatch blind-spot camera for the right lane.

Mendel attributed the forthcoming Civic's combination of strength and lightweight to what he termed "an innovative approach to body construction," though he gave no more details.

Although no specific measures were provided, the company did say that the new Civic rolls on a longer wheelbase and more overall length. Front overhang is reduced and the A-pillar moved rearward for a longer hood.

"Charismatic, connected, and athletic--our goal was to create a design that is emotion and sporty with a sense of timelessness and simplicity," explained Guy Melville-Brown, Exterior Lead Designer for the concept.

Though 20-inch wheels are a show-car staple, they seem an unlikely production option considering Honda's longstanding conservatism on wheel size.

The two-door coupe concept and the four-door sedan were designed by Honda's California studio and will be manufactured in the company's Alliston, Ontario, plant. The Greensburg, IN, factory will build only sedans. A differently styled five-door hatchback will come from the Swindon, England plant.

Honda confirmed that the previously announced turbocharged, direct-injected 1.5-L Earth Dreams four-cylinder engine will be the Civic's premium engine, while a naturally aspirated DOHC VTEC engine of undisclosed displacement will be the car's base engine.

Though 20-inch wheels are a show-car staple, they seem an unlikely production option considering Honda's longstanding conservatism on wheel size.

For North American vehicles, the turbo engine will be made in the Anna, OH, plant. Two transmissions will be available, a six-speed manual and a CVT that will be built at Honda's Russells Point, OH factory. The company says this new CVT will be sportier and more efficient than current examples.

"The last model didn't resonate with the target audience," noted Karl Brauer, Senior Analyst for Kelley Blue Book. "They've done everything they should have done to address their problems, with improved styling, a turbocharged engine, and multiple body styles."

Another key factor, he said, will be the arrival in North America of the racy Civic Type R, an ultimate performance model built on the five-door hatchback version from England. For the first time that car will come to the U.S., providing a halo car beyond the Civic Si coupe. That will let the Civic contend with rarified competitors like the Subaru WRX STI, Ford Focus RS and Volkswagen Golf R for enthusiasts' mindshare and dollars. ■

*Dan Carney*

# Honda Pilot Lighter, more powerful 3rd-gen Pilot debuts Honda's NPD process

**The available Honda Sensing suite of advanced safety and driver-assistance technologies, including a Collision Mitigation Braking System, helped the new Pilot earn the 2015 Top Safety Pick+ rating from IIHS.**

Programs for new re-engineered vehicles pose a great challenge for the engineers and other professionals assigned to them: load the vehicle with more advanced technology, consumer-friendly amenities, and safety features required to meet stringent requirements, all while trimming weight from any area necessary to maintain—or ideally decrease—overall weight compared to the previous generation. The 2016 Honda Pilot development team, faced with said challenge, succeeded brilliantly.

The new three-row SUV is 289 lb (131 kg) lighter—the previous Pilot Touring weighed 4632 lb (2101 kg) vs. the new Pilot Elite, which includes more features but weighs 4343 lb (1970 kg)—while recently earning a 2015 Top Safety Pick+ rating from the Insurance Institute for Highway Safety (IIHS).

The 2016 Pilot adopts the second generation of Honda's Advanced Compatibility Engineering (ACE) body structure along with a new "3-bone" underbody frame design and new hot-stamped ultra-high strength steel (UHSS) door rings.

Located under the front floor of the Pilot, the 3-bone structure improves impact load management, directing energy around the passenger cabin in the event of a frontal collision. The structure creates three different load pathways that channel collision energy. One channels collision forces from the front of the vehicle directly underneath the passenger cabin; the other two channel collision forces under the vehicle's left and right side frames. The structure saves 61 lb (28 kg).

According to Marc Ernst, Pilot Development Leader, the Pilot platform is more than 80% similar to that of the Acura MDX. And similar to the MDX, the new Pilot employs multiple lightweight, high-strength materials for its construction,

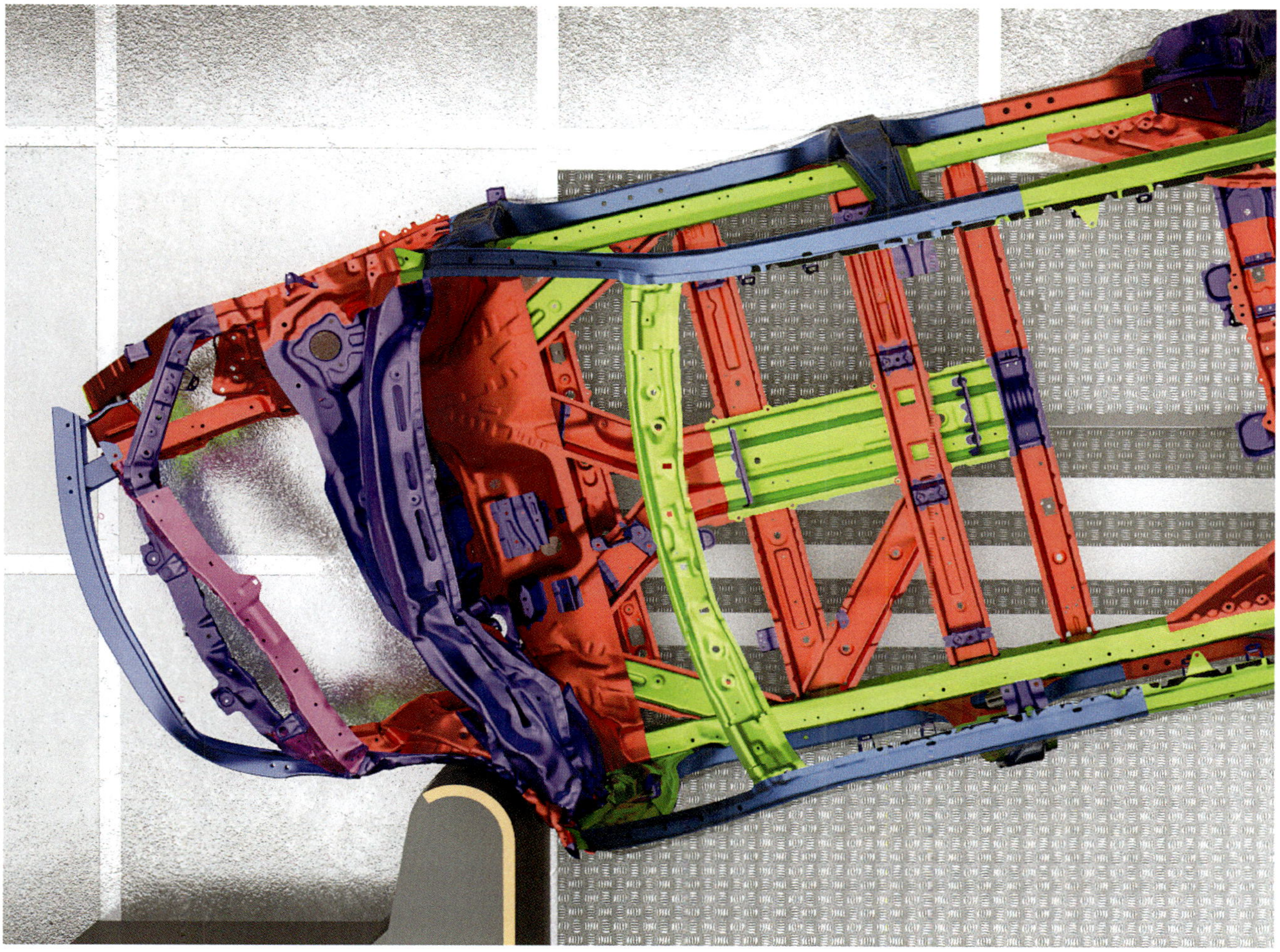

**The 2016 Pilot adopts the second generation of Honda's Advanced Compatibility Engineering (ACE) body structure along with a new "3-bone" underbody frame design and new hot-stamped ultra-high strength steel (1300 to 1500 MPa) door beams and rings.**

including seven different grades of steel; an aluminum hood and front bumper-reinforcement beam; and a cast-magnesium steering hanger beam. In total, 61.5% of the body structure is composed of advanced materials, most notably 21.3% of UHSS ranging from 980 to 1500 MPa, which is used for the front door outer stiffener rings.

In addition to enhanced crash protection, a result of this multi-material structure is 25% more torsional rigidity than the previous-generation SUV, as well as 67% better mount rigidity. Another benefit of using UHSS is improved visibility due to narrower A- and B-pillars. Total outward visibility for the driver improves from 271.4° on the previous model to 283.5° on the 2016 Pilot.

The SUV is 3.5 in (89 mm) longer, at 194.5 in (4940 mm), with 1.75 in (44 mm) of that increase in the wheelbase. A 1-in (25-mm) lower overall vehicle height and a subtly narrower width give a sleeker look and aid aerodynamics. Compared to the previous Pilot, specific aero refinements include a 1.93% reduction in frontal area, a full-width air dam, and underbody details such as air strakes and covers. CFD simulations show a 10% reduction in coefficient of drag (CdA) for the new model.

## Chassis and powertrain refinements

The Pilot's new front strut suspension includes the first use of Amplitude Reactive Dampers. Smaller driveshaft angles help to mitigate torque steer, and a reduced center offset reduces steering wheel vibrations. At the rear is a multi-link design with a tubular stabilizer bar. Compared to the previous Pilot, where the setup transmitted suspension loads to the vehicle's rear wheelhouse, the new setup transmits suspension loads to the more rigid body side frame, which allows for more aggressive tuning of suspension bushings, according to Ernst. The elimination of the forward trailing arm on the previous Pilot rear suspension allows for a 1.4-in (36-mm) lower rear floor.

A 16.0:1 steering ratio for the new Pilot is about 10% quicker than the previous model's steering, which complements the functionality of the torque-vectoring rear axle on AWD models. The available new Intelligent Variable Torque Management (i-VTM4) AWD system with an electronically controlled rear differential and dynamic torque vectoring capabilities aids both traction and cornering performance. Other benefits include 46% faster all-wheel response time and 20% greater torque capacity. The new system, which is Honda's first

**The new Pilot features premium interior upgrades, including available heated and ventilated front seats, a heated steering wheel, heated second-row seats, and a Honda-first panoramic glass roof option—all offered as standard equipment on the new top-of-the-line Pilot Elite trim.**

U.S. application of torque vectoring, is 22 lb (10 kg) lighter than the previous one, with reduced drag.

Powering the 2016 Pilot is a 3.5-L direct-injected i-VTEC V6 engine with Variable Cylinder Management cylinder deactivation. The new engine is mated to one of two new transmissions: a 6-speed automatic (6AT) for LX, EX, and EX-L trims, and 9-speed automatic (9AT) for Touring and Elite models. Models with the 9AT also have a new Idle-Stop capability to enhance fuel efficiency.

The EarthDreams series V6 is rated at 280 peak hp (209 kW) and 262 lb·ft (355 N·m) peak torque—increases of 30 hp (22 kW) and 9 lb·ft (12 N·m) over the previous model. EPA fuel-economy ratings (city/highway/combined) are up across all trim levels. The 2WD model with 9AT gets 20/27/23 mpg.

The new Pilot was designed and developed by Honda R&D Americas, Inc., in Los Angeles and Ohio, and is manufactured by Honda Mfg. Alabama, Inc. in Lincoln, AL.

## Honda's NPD process

With nine trim levels and a high level of feature content, the 2016 Pilot could have been a daunting challenge for Honda associates who assemble the all-new vehicle in Lincoln, AL, as well as for the program's systems suppliers.

"All this [new] content created a tremendous amount of build complexity," noted Jeff Tomko, President of Honda of Alabama Manufacturing. He said during early planning discussions Honda R&D engineers realized the need for more flexible processes that made the larger vehicles easier to assemble without sacrificing quality.

The solution was NPD (No Prototype Development), a digital product development-and-manufacturing process that makes its full-scale debut on the new Pilot. NPD aims to ultimately eliminate all physical prototypes prior to their requirement in final vehicle-crash certification. The "virtual build"

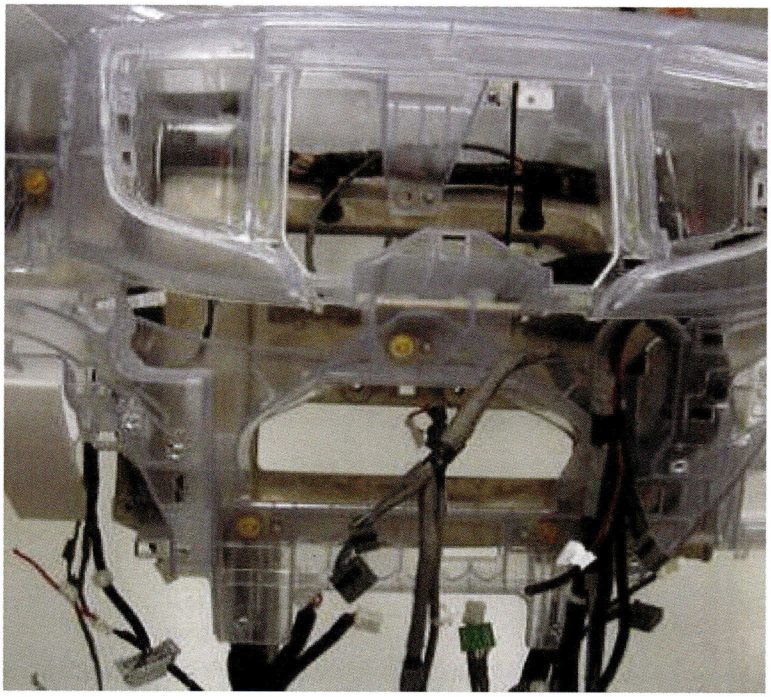

**A 3D-printed transparent IP cover greatly aided design of the Pilot's internal wiring.**

concept has long been an industry aspiration, of course, and Honda has been working up its NPD systematically, employing it first on some aspects of the 2015 Acura TLX sedan program.

"The key to NPD is close collaboration between our R&D engineers and manufacturing teams," Tomko told *Automotive Engineering*. Playing a vital supporting role is Dassault Systemes' comprehensive suite of 3D applications—CATIA (design/engineering), DELMIA (manufacturing operations), SIMULIA (FEA and simulation), and 3DEXCITE (real-time visualization)—that are used by Honda. Working for the first time without hard prototypes—previously R&D would build about 100 white bodies per program, for example—the R&D-and-plant-floor teams employed a new training strategy, with each department establishing a "Top Gun" go-to expert for rapid leveraging of know-how.

As a result, design engineers gained first-hand experience with Pilot production challenges during early trials in the plant, rather than at the R&D center. An important time-saver was Honda's expanded use of 3D printing to support process development. Examples included creation of a special line-side ergonomic lifting device for installing the Pilot's new electric power-steering motor unit; a transparent IP cover that served as a troubleshooting tool; and a center console 3D-printed to determine ease of assembly. ■

*Ryan Gehm and Lindsay Brooke*

# Redesigned, re-engineered 2016 Honda Pilot

**The 2016 Honda Pilot will offer 20-in wheels. "It's a first-time application for any Honda model as a non-accessory from the aftermarket," said Honda's Rudy Mayne.**

From its panoramic roof to its platform, the third-generation Honda Pilot is vastly different from its predecessor.

"From top to bottom we've advanced every aspect of the new Pilot, appealing to buyers' diverse needs and wants," said Marc Ernst, the 2016 Honda Pilot's Chief Engineer and Large Project Manager.

The SUV made its world premiere at the 2015 Chicago Auto Show.

The three-row midsize utility vehicle's optional panoramic roof is a new Pilot feature. "We're able to maintain our roof-mounted rear entertainment system, which is the best place for viewing, especially for the third row," Rudy Mayne, Honda's Assistant Large Project Leader, said in an interview with *Automotive Engineering*. The fixed-glass panoramic roof with power shade complements a separate sunroof for front-row occupants.

Honda's all-new global midsize light-truck platform debuted on the 2014 Acura MDX luxury SUV.

"The 2016 Pilot is the second vehicle off that platform, so there is some commonality in the basic platform," Ernst told *Automotive Engineering*. "Moving forward, this platform will be the basis for the Ridgeline. Of course, there will be some different tuning because it's a pickup truck. And this platform will be the basis for the next Odyssey [minivan]."

The 2016 Pilot's rear differential system is lighter weight and more efficient than the former viscous coupling design.

"This new electronically controlled system gives us a lot of advantages in terms of putting the torque where it really needs to be and sensing wheel slip before the actual slip," said Ernst. This is a first application for Honda. The Acura MDX has a similar system, but it is tuned differently.

Ernst said the 2016 Pilot will be powered by the latest version of the 3.5-L direct-injected i-VTEC V6 engine with Variable Cylinder Management cylinder deactivation technology. Replacing the 2015 Pilot's five-speed automatic transmission are two new transmissions: a Honda-developed six-speed automatic transmission and a ZF-developed nine-speed automatic transmission. The six-speed is a Pilot-first application, while the nine-speed is a Honda-first application.

Honda officials are not yet releasing horsepower, torque, or fuel-economy ratings, but Ernst said the new powertrain, the approximate 300-lb (136-kg) drop in vehicle weight, and the anticipated best-in-class aerodynamics should equate to best-in-class fuel economy. The 2WD 2015 Pilot's mpg is 18 city/25 highway/21 combined, while the 4WD model's mpg is 17 city/24 highway/20 combined.

**Third-generation Honda Pilot's technology offerings include remote engine start, idle stop, heated and ventilated front seats, heated steering wheel, and heated second-row seats.**

SAE Eye on Engineering Quick Take:

## Honda Pilot 5-star safety

Watch the video at https://youtu.be/zVWDD5kcOSI

For automakers, earning a 5-star safety rating in the government and insurance industry tests is a major source of pride. In this episode of SAE *Eye on Engineering*, Senior Editor Lindsay Brooke looks at the latest vehicle to earn 5 stars: Honda's 2016 Pilot. SAE *Eye on Engineering* can be viewed at https://youtu.be/zVWDD5kcOSI. It also airs in audio-only form Monday mornings on WJR 760 AM Detroit's Paul W. Smith Show.

Access archived episodes of SAE *Eye on Engineering* at www.sae.org/magazines/podcasts.

SAE Eye on Engineering Quick Take:

## Honda's U.S. engineering

Watch the video at https://youtu.be/fwarDAh5lGU

More and more, Honda vehicles are getting their start on the CAD screens of American designers based in central Ohio, rather than those in Japan. In this episode of SAE *Eye on Engineering*, Senior Editor Lindsay Brooke looks at the busy year for Honda engineers in Raymond, OH. SAE *Eye on Engineering* can be viewed at https://youtu.be/fwarDAh5lGU. It also airs in audio-only form Monday mornings on WJR 760 AM Detroit's Paul W. Smith Show.

Access archived episodes of SAE Eye on Engineering at www.sae.org/magazines/podcasts.

With 3 in (76 mm) of added vehicle length, the new Pilot has a larger cargo area than the 2015 model. Entry and exit from third-row seating improves with a 1-in (25-mm) lower step-in height. An available one-touch button conveys an automatic forward slide/forward tilt of second-row seating. The second row is a standard 60/40 split bench seat, or a Honda-first application of optional captain's chairs.

Pricing will be announced closer to the 2016 Pilot's launch at Honda dealerships in summer 2015. All three generations of Pilot were designed and developed by Honda R&D Americas Inc. in California and Ohio. The 2016 Pilot and its engine will be assembled at the automaker's Lincoln, AL, plant. ■

*Kami Buchholz*

# Honda enters burgeoning subcompact SUV fray

**The Honda HR-V aims to let drivers combine tidy, easy-to-park dimensions with a spacious passenger-carrying back seat and an affordable price tag.**

Size creep in the compact crossover segment has seen Honda's CR-V outgrow some of its original target customers. So, as it did when the Fit slid into the product lineup beneath the Civic, Honda has added the sub-compact HR-V to its crossover family beneath the enlarged CR-V.

This new crossover is close to the size, weight, and interior volume of the first-generation CR-V, but with slightly more horsepower, comparable torque, and a wider spread of gear ratios, so performance and efficiency are much better than that older model. Naturally, safety is hugely improved as well.

The HR-V is built on Honda's Global Compact Series platform, with its center fuel tank and second-row fold-in-floor Magic Seat as with the Fit. Compared to a Fit, the HR-V is 9.1 in (231 mm) longer, 2.8 in (71 mm) wider, 3.2 in (81 mm) taller, and rolls on a 3.2 in longer wheelbase.

The HR-V enjoys many upgrades in comparison to the Fit, because its additional size requires more power and because the company expects the HR-V's customers to be a bit more upscale, with greater demands for features than Fit buyers.

"If we had used only Fit technology in the HR-V, that would have been boring," observed Naohisa Morishita, HR-V Chief Engineer and Development Leader. "We've taken technology from higher-class vehicles to provide value for this new segment."

The HR-V retains the Fit's MacPherson strut front and torsion beam rear suspension layouts, but the struts are Sachs amplitude reactive dampers that provide separate tuning for large-displacement compression and rebound events. This helps the HR-V ride more smoothly, while preserving the ability for its dampers to stiffen as needed to control big jolts.

The HR-V borrows the RealTime all-wheel drive system from the larger, heavier CR-V. Though the hardware is unchanged, its performance is improved by its application in the lighter, less-powerful HR-V, said Morishita. The maximum rear torque split remains 50%, but it will tend to provide more power to the rear more of the time than the CR-V's system, he explained. The difference won't likely be noticeable on dry pavement, but in slippery conditions the HR-V will feel more responsive, Morishita said.

Because the system uses the HR-V's various sensors to determine that it is driving in snow, rather than simply waiting to react to wheelspin, it will automatically send more power to the rear wheels when starting from a stop in snow to reduce the chance of front wheel spin.

The HR-V's 141-hp (105-kW), 127-lb·ft (172-N·m) SOHC 1.8-L four-cylinder engine comes directly from the Civic, giving the HR-V a needed power boost over the Fit. For the HR-V application, torque is optimized in the 2000-4000 rpm range, with the variable valve timing system changing from one intake valve to two for the 1000-3000 rpm range, depending upon throttle position and load. Similarly, a shutter valve in the composite dual-stage intake manifold switches from torque-fortifying long intake runners to high-flow short runners at 5000 rpm for peak power.

Available transmissions are a CVT automatic that boasts a wider spread of ratios than that in the CR-V, covering a range from 2.526: to 0.408:1. The six-speed manual transmission, which lends the HR-V a livelier feel, is available only with front-wheel drive; the CVT can be matched to either the standard front-wheel drive or the available all-wheel drive.

U.S. EPA fuel economy ranges from 25 mpg city and 34 mpg highway for the front-drive manual transmission model to 28/35 for the front-drive CVT version. The all-wheel-drive CVT scores 27/32.

Structurally, Honda's engineers went to great lengths to reinforce the HR-V's unibody. "One of the reasons is to improve its handling and dynamic performance," noted Morishita. It also contributes to Honda's expectation of maximum ratings on U.S. NHTSA and IIHS crash tests.

The HR-V's body shell includes 27% extremely high-strength steel of 780 MPa (113 ksi) or higher, with the hardest 1500-MPa (218-ksi) steel deployed in the door sill, A-pillar, and roofline over the front door opening. The company also reinforced the attachment points for the roof pillars to add strength at the joints, but some of that improvement comes from increased material at the joint.

The HR-V's rear hatch opening also received special attention. "It is a two-box design," he noted, so "it can become torsionally weak. We wanted to improve on that." To strengthen the hatch area, the HR-V uses a single stamping of high-strength steel to ring the door opening.

Inside, "it has space for your dreams, and when you put your foot down, you feel like you've made it," Morishita enthused. The cabin can be configured in normal mode, with regular seating for five; in split mode, with space for a long item in the rear and one back-seat passenger; tall mode, with the rear seat bottoms flipped up against the seat backs to clear space from the floor to the roof; utility mode, which sees the rear seats folded flat for a larger cargo bay; and long mode, when the front passenger's seatback is also folded down to permit carriage of very long objects entirely inside the HR-V.

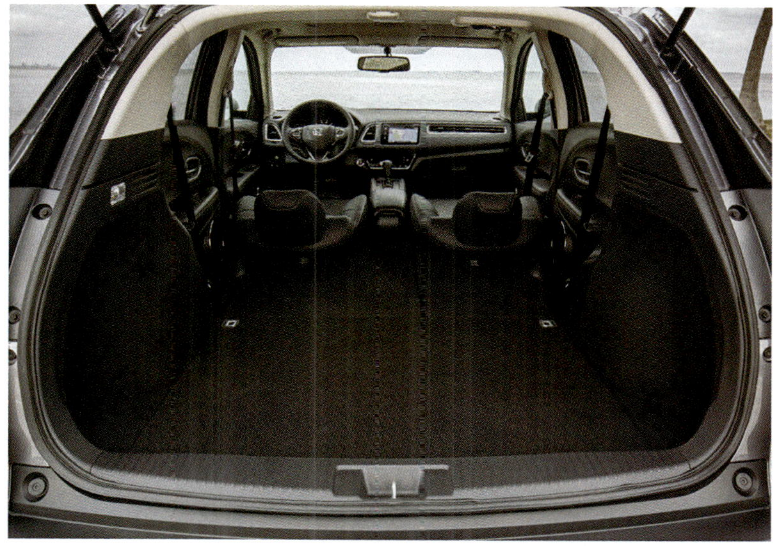

**Honda engineers reinforced the rear hatch opening with high-strength steel to retain structural rigidity while providing an expansive aperture for cargo.**

**U.S.-spec HR-Vs will enjoy power from the 141-hp 1.8-L engine borrowed from the Civic.**

The HR-V comes with a high level of standard equipment, including exterior cues such as its 17-in polished aluminum wheels and cabin technology like a back-up camera. The company said it thinks this combination of affordability and amenities will attract 70,000 annual sales for the HR-V in the U.S. alone. ■

*Dan Carney*

# Hyundai enhances NVH with 2016 Tucson's re-engineered suspension—including 'world first' dual-member damper housing

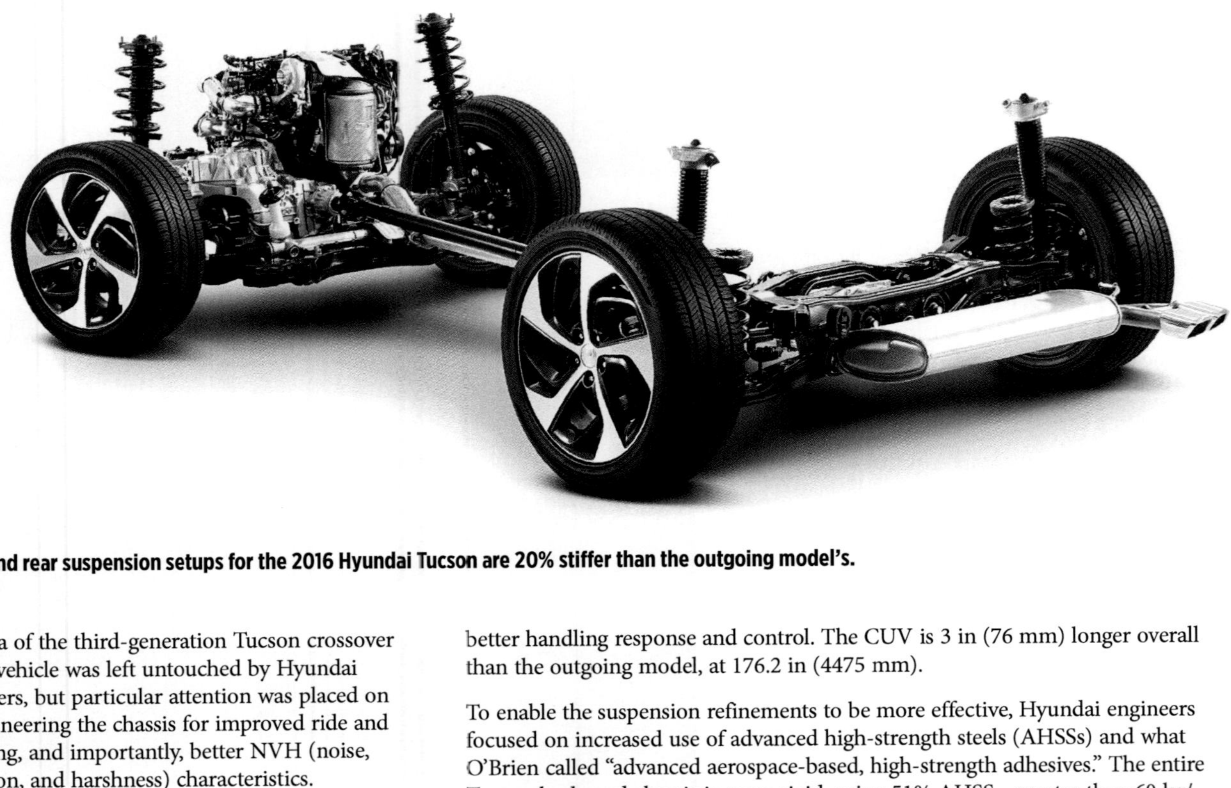

**Front and rear suspension setups for the 2016 Hyundai Tucson are 20% stiffer than the outgoing model's.**

No area of the third-generation Tucson crossover utility vehicle was left untouched by Hyundai engineers, but particular attention was placed on re-engineering the chassis for improved ride and handling, and importantly, better NVH (noise, vibration, and harshness) characteristics.

Not surprisingly, the 2016 Tucson platform is larger than the previous generation's; however, engineers focused on improving width and wheelbase for greater ride comfort and linear stability, according to Mike O'Brien, Vice President, Corporate and Product Planning, Hyundai Motor America (HMA).

"We have a full lineup of CUVs, so having the biggest one in the segment is not the best because people that buy in this area of the market focus more on maneuverability and parking," O'Brien said. "So instead we focused on getting the best width that we could, at 72.8 in, and also the best wheelbase at 105.1 in, which gives a sense of stability."

Compared to the 2015 model, the new Tucson is 1.1 in (27.9 mm) wider, with a 1.2-in (30.5-mm) increase in wheelbase. The front and rear overhangs were increased "in smaller proportion," centering more of the Tucson's mass within the wheelbase for better handling response and control. The CUV is 3 in (76 mm) longer overall than the outgoing model, at 176.2 in (4475 mm).

To enable the suspension refinements to be more effective, Hyundai engineers focused on increased use of advanced high-strength steels (AHSSs) and what O'Brien called "advanced aerospace-based, high-strength adhesives." The entire Tucson body and chassis is more rigid, using 51% AHSS—greater than 60 kg/mm2 (590 MPa)—compared with 18% for the former model.

"Another thing that we've done, very similar to what we did with Genesis, is to use a tremendous amount of high-strength adhesives," said O'Brien. "We went from no adhesives—just a welded structure on our outgoing model—to now over a football field in length of adhesives [335 ft (102 m), to be exact]. Not only does this help in body strength, but also in NVH by isolating noise better than just welding."

The result of such efforts is a 48% improvement in body rigidity.

## 'World first' mounting structure

One area that benefits from the employment of AHSS is a "world's first" shock-absorber mounting structure for the 2016 Tucson. The CUV employs a dual-reinforcing panel rear wheelhouse design, which optimizes panels that are prone to vibration, resulting in a 109% increase in rigidity, reduced road noise levels, and ride and handling improvements. Traditional designs employ only one reinforcement.

**Compared to the 2015 model, the new Tucson is 1.1 in (27.9 mm) wider, with a 1.2-in (30.5-mm) increase in wheelbase.**

"Hot-stamped steel is the highest stiffness steel we use. Generally it is used for crash [protection], but we also use it for the [integrated dual-member rear wheelhouse] structure," said Chahe Apelian, Senior Manager of Vehicle Evaluation – Chassis & NVH, Hyundai America Technical Center, Inc., as he gave *Automotive Engineering* a walk-around of the Tucson body-in-white (BIW) cutaway.

"In terms of the rear suspension towers, it's to make the attachments stiff enough to where we increase the envelope of tunability of the shock absorbers," he continued. "Once the body becomes stiff to the point where it deflects very little, all the energy from the road is absorbed by the shock absorber and that's a tunable variable."

Apelian pointed out the spring seat on the underside of the BIW wheel well, comprised of 100 kg/mm2 (980 MPa) HSS and hot-stamped steel; "on the other side, we sandwich it with another piece of hot-stamped steel, so it's all bracketed in and takes all the forces right here."

Engineers in Korea began working on the solution since the previous generation, according to Apelian, after identifying the rear suspension mount as an area that needed improvement. HSS being a "core technology" for Hyundai, its employment for this application made sense, he said.

"The part sizes aren't very big; it's not like we're using a whole rocker," he explained. "It's a laser-focused application, so it minimizes cost but gives us the biggest bang for the buck.. You could do it with normal steel, but then you start adding more weight."

Though the overall solution is heavier than the previous single reinforcement, Apelian noted that "all of the other solutions would be higher weight…to achieve this much stiffness and this much ride improvement."

## Other NVH enhancements

Another area targeted for NVH improvements was the subframes, which are fully isolated, front and rear. "The platform is basically a derivative of the Sonata platform, and we carried over those," he said. "The body stiffness is critical to making sure those subframes work, so that the handling doesn't degrade, and also for managing the 19-in tires, making sure they work with the whole system."

The 19-in tires are one size larger than the previous-generation's available 18-in tires.

Front suspension is a MacPherson strut design, with coil springs, gas-filled shock absorbers, and a 24.7-mm (0.97-in) stabilizer bar. The rear suspension is an independent, multi-link design, now with dual lower arms for both front-wheel and

**The new Tucson features a "world first" integrated dual-member rear wheelhouse design using AHSS (orange = 100 kg/mm2, or 980 MPa; red = hot-stamped steel). Also shown is the hydraulic rebound bump stop.**

The spring seat on the underside of the body-in-white (BIW) wheel well.

Hyundai's Chahe Apelian next to the Tucson BIW. Note the hot-stamped steel piece (red) that sandwiches the damper, resulting in a 109% increase in rigidity, reduced road noise levels, and ride and handling improvements.

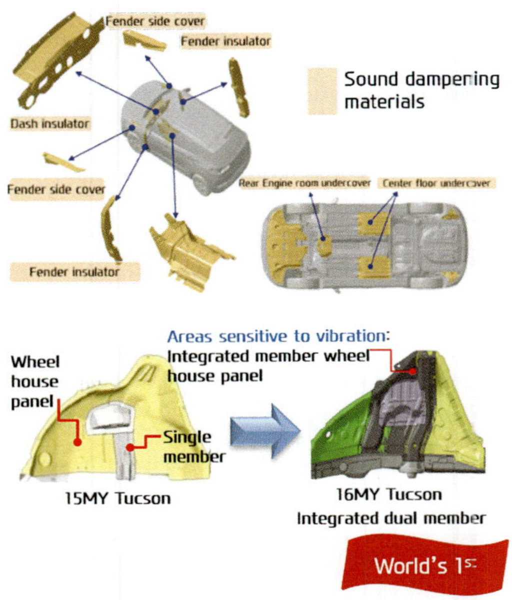


| Item (dBA) | Tucson | RAV4 | CR-V | Escape |
|---|---|---|---|---|
| Wind Noise | **67.2** | 68.6 | 70.2 | 67.9 |
| Road Noise | **65.5** | 66.4 | 68.6 | 68.3 |
| Idle Noise | **61.7** | 62.7 | 65.0 | 62.9 |

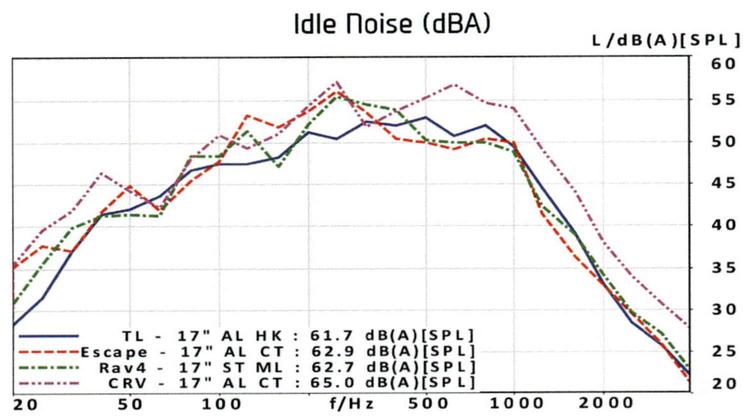


Hyundai Motor America

HMA Internal Competitive Evaluation; Wind Noise (Smooth Road @ 110 kph) / Road Noise (Coarse Road @ 60 kph)

**Hyundai engineers' focus on chassis and NVH enhancements paid off in internal testing of noise levels.**

all-wheel-drive models, for optimal ride comfort and body control. Overall, front and rear suspension setups are now 20% stiffer than the outgoing Tucson's.

Suspension bushings are now composed of a higher-damping synthetic rubber that is 30% stiffer for smoother ride characteristics when compared with conventional rubber.

"The front suspension was completely redesigned," O'Brien said. "We went to a four-point bushing setup for better rigidity and better noise isolation. Probably more importantly we've gone from a conventional rubber bump stop that basically has that very harsh feeling when you hit the limit of the suspension travel, to instead a hydraulic rebound spring stopper that manages that last bit of suspension travel much more comfortably."

To help lower wind noise, aerodynamics have been improved to 0.33 Cd, a 0.02 lower coefficient of drag than the former Tucson.

Based on Hyundai's internal tests at its R&D center in Ann Arbor, MI, the new Tucson has achieved better road noise (65.5 dBA), idle noise (61.7 dBA), and wind noise (67.2 dBA) measures than Toyota RAV-4, Honda CR-V, and Ford Escape (see image above)—"the result of our extra effort in terms of NVH and road noise isolation," said O'Brien. ■

*Ryan Gehm*

# Hyundai updates its Tucson compact crossover

**Hyundai's European design team, led by Thomas Buerkle, has worked hard to give the new Tucson an improved presence on the road with a more dominant grille treatment.**

The all-new Tucson shown at the 2015 Geneva Motor Show represents a major step forward for Hyundai in Europe, where the C-SUV segment accounted for 22% of the company's sales in 2014.

As well as sitting on a new platform and being both longer and wider than the outgoing model, it represents a new styling theme according to Hyundai Motor Europe's Chief Designer, Thomas Buerkle.

"We call it 'Fluidic Sculpture,'" he said. "Its execution is purer and simpler than the previous car, the sculpture being emphasized on the complete body rather than just on features. So, for example, the rear fender is sculpted with a striking feature line that starts just above the front wheel arch and rises rearwards over the taillights.

"We made the hexagon grille more visible with a thick chrome frame; we wanted to make it more pure and focused. The lamps with LEDs have this sleek, high-tech look that gives the car a wider stance. Whilst the lower flying bar as we call it, with the day light running lamps adds to the car's width to give it a good stance. What was very important for us was that we pushed the 'A' pillar [in] to increase the bonnet's length, this not only improves visibility but adds to the car's prestige."

Buerkle says the car is "more serious, more adult" than the previous car adding that Hyundai wanted to reduce the number of external features to give the Tucson a stronger image: "What is very characteristic are the asymmetric wheel arches that have a forward moving element giving the car a more dynamic stance."

Inside there has been a conscious upgrade of materials, fit, and finish, said Buerkle, with a more horizontal structure to the fascia; the central monitor is now level with the instrument cluster for improved visibility. "We wanted to make the interior sleeker with a wing profile," Buerkle explained, "to make it feel bigger and more modern looking."

"Materials were a key issue for the color and trim team, the two tones make the interior feel more sleek, whilst the leather's tactility and the feel of the switches have been greatly improved and upgraded," he added.

Technically, there's a new electric-motor-driven power steering (MDPS) system featuring newly developed software and a rack-mounted electric motor for fast and direct response.

Based on an all-new platform, the Tucson is both longer and wider than the outgoing model with a increased wheelbase for additional interior space.

**Improved interior layout now features a wing profile for the fascia with the central monitor raised to the same level as the instrument cluster for improved clarity.**

The new Drive Mode Select (DMS) function on automatics offers customers a choice of two drive modes—Normal and Sport—with different characteristics for the steering, throttle, engine, and ratios.

The Tucson will be available with one of the widest powertrain ranges in its class. These include two gasoline engines plus three diesel options.

The gasoline units are the 130-hp (97-kW) 1.6-L GDI or 172-hp (128-kW) 1.6-L T-GDI. The new turbocharged 1.6-L T-GDI gasoline engine with 172 hp delivers maximum torque of 195 lb·ft (264 N·m) from 1500 to 4500 rpm, and it is offered with a six-speed manual or the seven-speed dual-clutch transmission (7DCT).

The diesel options are a 113-hp (84-kW) 1.7-L, 134-hp (100-kW) 2.0-L standard power, or 183-hp (136-kW) 2.0-L high power. The "low" power 113-hp diesel engine delivers maximum torque of 206 lb·ft (279 N·m) from 1250 to 2500 rpm and is available exclusively with a six-speed manual transmission, 2WD, and fuel-saving stop-start technology. The 2.0-L turbodiesel standard power delivers 275 lb·ft (373 N·m) of maximum torque at 1500 to 2500 rpm. The high power version delivers an impressive max torque of 296 lb·ft (401 N·m) from 1750 to 2750 rpm. Both feature maintenance-free lean NOx trap catalyst technology to reduce emissions.

The 1.6-L T-GDI petrol and 2.0-L diesel engines can be mated with a 4WD-system, which delivers 100% of torque to the front wheels during normal road driving and up to 40% to the rear wheels, automatically, depending on conditions. A manually-selected Lock Mode splits torque 50/50% for enhanced stability at low speeds.

All engines have been enhanced to offer improved driveability and fuel efficiency, and they meet Euro6 emissions standards; stop-start is available on selected engines.

New convenience features include a smart powered tailgate that opens as the key-holder approaches; smart parking assist system with parallel and bay parking functions; autonomous emergency braking; lane keep assistance; rear cross traffic alert; and an active hood system that raises the bonnet if impacted by a pedestrian or cyclist. ■

*Ian Adcock*

# Jaguar enters performance crossover SUV segment

**Jaguar Design Director Ian Callum introduces the F-Pace in Frankfurt.**

In a bid to combat flagging market share due to declining consumer interest in its signature coupes, roadsters, and sedans, Jaguar is making its first move into the booming market for crossover SUVs. The 2016 F-Pace, shown at the 2015 Frankfurt Motor Show, is an aluminum-intensive "performance crossover," Jaguar's term for an all-weather, on-road-focused wagon with real handling on pavement and no pretensions that it is meant to conquer the Rubicon Trail.

Philosophically, this is similar to Porsche's wildly popular Macan, a vehicle Jaguar used as a benchmark for the F-Pace. Jaguar calls its foundation the Lightweight Aluminum Architecture, which it shares with the XE and XF sedans. "We recognize that the dynamic benchmark in the segment is the Porsche Macan," acknowledged Vehicle Program Director Andy Whyman.

The company shies from the term "platform" with its implicit parts sharing because the F-Pace is 81% unique from its sedan siblings. The result is a vehicle that is 80% aluminum by weight, which is more than any other crossover SUV, said Kevin Stride, Jaguar XE Vehicle Line Director. Areas that are normally high-strength steel for its strength in crashes are high-strength aluminum in the F-Pace, including the B-pillar and the door-surround stamping. Shock towers are high-pressure die-cast aluminum.

**The F-Pace is more than 80% aluminum by weight even using aluminum for parts like suspension castings and the B-pillar.**

Jaguar boasts that one-third of the aluminum used in the F-Pace is recycled, and the goal is to raise this portion to 75% by 2020. This currently amounts to 30,000 ton (27,200 t) of reused aluminum annually. It should be understood that this refers to the company's reuse of its own scrap aluminum from its stamping

**The first F-Pace vehicles to come to the U.S. will be powered by Jaguar's 2.0-L Ingenium turbodiesel four-cylinder engine.**

facility, and not to the notion that discarded beer cans have found a higher purpose.

That's because most of the aluminum used in the F-Pace is a very specific RC5754 grade developed with Novelis for its strength properties. "We are working on using post-consumer material," Whyman said. Floorpan stampings are the most likely application for such recycled metals, he said. "Beer cans are similar to floorpan material," he explained. The structural aluminum is all high-strength.

Other innovative materials include the use of magnesium for a front frame crossmember as well as the common dashboard crossmember. And the vehicle's rear hatch is composite. The resulting body-in-white weighs 298 kg (657 lb), a mass comparable to that of the much smaller Fiat 500L's.

With 100,000 h of computer simulation time and 15,000 h of engineering work in the F-Pace's structure, Jaguar has been able to optimize its design to produce a vehicle that is 50% stiffer than the Macan in lateral stiffness at the front end and 35% stiffer across the rear, according to Whyman.

This stiffness, along with Jaguar's own software for the Bosch Automotive Steering-sourced variable-ratio electric power steering system contributes to the F-Pace's crisp steering response, he said.

Front suspension is double-wishbone and the rear is what Jaguar terms an "integral link" double-wishbone design whose castor properties contribute to stability during hard braking, said Whyman. It also provides increased longitudinal compliance to cushion the blow from sharp-edged impacts such as potholes. Dampers are Bilstein units. Another aluminum application is for the front steering knuckles, which are forged from cast aluminum blanks for maximum stiffness.

At launch the F-Pace will offer five engines, three of which will come to the U.S. market; a 180-PS (132-kW) 2.0-L turbodiesel four-cylinder, a 340-PS (250-kW) 3.0-L supercharged gasoline V6, and a 380-PS (270-kW) version of the same engine.

The Ingenium diesel engine employs variable exhaust cam timing to quickly heat the exhaust aftertreatment system on cold starts. The engine itself also warms quickly thanks to a split cooling system, variable water pump, and a mapped thermostat. The 1800-bar (26.1-ksi) common-rail injection system and variable-ratio turbocharger are complemented by a cooled low-pressure EGR system to produce minimal NOx.

The gasoline V6 engines are the same as seen in the F-Type sports car, but tuned for more tractable low-rpm response. World markets also get a 300-PS (220-kW) 3.0-L twin-turbocharged diesel V6 and a turbocharged 2.0-L gasoline four-cylinder engine of unspecified output. Company sources indicate that the gasoline four-cylinder will come to the U.S. later.

All engines are matched to ZF eight-speed automatic transmissions, though a six-speed manual

**The BorgWarner all-wheel-drive transfer case adds only 8 kg to the F-Pace over the 2WD variant. Dan Carney**

**Jaguar worked with SuperAlloy on the wheels and Pirelli on the tires to ensure that the F-Pace's wheels have some protection against curb scrapes. Dan Carney**

transmission is available with the diesel four-cylinder in some markets. Rear-wheel drive is standard for the diesel, with optional all-wheel drive and the gasoline V6 models are all-wheel drive only.

The F-Pace's BorgWarner all-wheel-drive system is an evolution of that seen on other Jaguar models, one that sees the weight trimmed by 16% and internal losses reduced by 10%. All-wheel-drive models weigh only 8 kg (18 lb) more than rear-drive ones.

Power reaches the road through wheels that are available in sizes between 18 and 22 in. The SuperAlloy Industrial Co. Ltd. 22-inch wheels wear tires with higher sidewalls than those on competitors' vehicles, helping shield them from increasingly common impact damage.

New technology abounds inside the F-Pace, where there is a 10.2-in touchscreen infotainment display and a 12.3-in instrument panel that are powered by an Intel quad-core processor with a 60-GB solid-state hard drive. This promises quick, responsive pinching, swiping and scrolling, but Jaguar has retained a familiar rotary knob for volume control for the 17-speaker, 825-W sound system, unlike the interface in Cadillac and Honda models that rely exclusively on touch surface inputs and steering wheel controls.

All of the system's components are connected via gigabit Ethernet networking, making the F-Pace one of the very first vehicles to employ this technology. This seems appropriate because the F-Pace has a built-in cellular Wifi capability using a roof-mounted antenna for 30% better reception than that of a handheld cell phone, the company said. It supports as many as eight connected wireless devices at a time.

This cellular connection also lets owners connect to their car remotely using Apple iOS or Google Android smart devices to check whether the doors are locked, whether any windows are open, where the F-Pace was last parked, and the levels of fuel and windshield washer fluid. They can use the app to lock or unlock the F-Pace, start the engine, receive alerts if the alarm is triggered, and reset it remotely.

An innovative new system called Active Key is a Flextronics-supplied rubber wristband RFID system that the driver wears like a Fitbit activity monitor. With it, the driver can lock keys and valuables in the car and unlock it by pressing the Active Key to the letter "J" in the F-Pace's rear Jaguar badge. A Delphi onboard system communicates with the passive Active Key, so there is no battery in the wristband to discharge.

Four-cylinder diesel F-Paces reach U.S. dealers this fall, with gasoline six-cylinders arriving in 2016. ■

*Dan Carney*

# Jaguar's making of a sports sedan

**It may be a standard premium sedan, but Jaguar chose to demonstrate its new XF to *Automotive Engineering* on the Spanish Navarra race circuit. This is the V6S.**

"Sport" is a much abused word in the automotive industry. Like "coupe" and "GT" together with other attempts at glorification through categorization, it is open to abuse.

So when Jaguar's senior engineering executives wanted to demonstrate to *Automotive Engineering* that its new, mid-size, lightened, aerodynamically efficient, premium 2016MY XF really is a "sports sedan," they illustrated the point not only on very demanding roads in northern Spain but also on the fast Circuito de Navarra racetrack.

With the supercharged 279-kW (374-hp) 460-N·m (339-lb·ft) 4.0-L V6 engine from the F-type tucked under its hood, the XF was able to show how to convincingly link, at high speed, the words "sedan" and "sport" even with this editor (with no race driver pretensions) in control. It tackled the task with a composed and tidy aplomb.

But the new XF (previously described at http://articles.sae.org/14038/) is also available with Jaguar's new 2.0-L four-cylinder Ingenium diesel—using Bosch 1800-bar (26.1-ksi) common-rail injection system—with power outputs from 120 to 132 kW (161 to 177 hp), the latter offered in "R-Sport" configuration.

In engineering terms could that, despite its nametag, also be termed a sports sedan? The answer, with one caveat, is in the affirmative.

## "Must" list

When Jaguar set out to create the new car, the "must" list of criteria that its engineers had to meet may have seemed daunting, but the result, explained Vehicle Engineering Director Ian Hoban, and Senior Manager Engine Planning and Strategy John Pepperell, is in line with expectations.

Compared to the outgoing XF introduced in 2007, these include a weight loss of up to 190 kg (419 lb) thanks to an aluminum-intensive architecture; torsional stiffness up by 28%; aerodynamic efficiency of 0.26 Cd; improved packaging on a 51 mm (2.0 in) longer wheelbase but 7 mm (0.3 in) shorter overall length; and external and internal design that marks it out as a premium product. The mechanical side includes integral link rear suspension; two- and all-wheel-drive versions; EPAS (electrical power assisted steering); and a very broad engine spread—there is also a 3.0-L V6 diesel with maximum torque of 700 N·m (516 lb·ft).

Of the car's 3194 part numbers, 2669 are new.

**A 75% aluminum intensive architecture is a major contributor to the new-generation Jaguar XF's agility.**

"The new XF shares its fundamental chassis aspects with the XE," said Hoban. "Double wishbones front and aluminum integral link rear suspension we believe are the best combination for handling and premium-level ride comfort."

Jaguar has accumulated 15 years' production experience of aluminum throughout its vehicles. "We work at the limit of what is feasible in regard to design and manufacture using the material," said Hoban. "Aluminum skin panels are used for A surfaces, aluminum extrusions for sills, and aluminum castings of varying thicknesses for the foundations of the car such as its front suspension mounts. Technologically challenging aluminum bodysides are single panels, each weighing less than 6 kg."

High-strength magnesium castings are used for the XF's front-end carrier and cross-car beam, with ultra-high-strength boron reinforcement, in particular its B-posts for side-impact protection.

Pepperell begins his involvement with all the company's new powertrains at a very early stage—as he did with those for the XF. That involvement concerns the embryo engineering programs in terms of production and manufacturing strategy.

"It includes technical target setting for the engines in terms of power, torque, refinement, economy,

**As Engine Planning and Strategy Senior Manager, John Pepperell begins his involvement with all new JLR engines at a very early stage.**

**The 2.0-L diesel version of the Jaguar XF can be had in R-Sport specification.**

Packed in tight: the supercharged V6 engine of the new XF is shared with the F-Type sports car.

Jaguar also used challenging northern Spanish roads to demonstrate its sports cars and sports sedans including the new XF.

emissions—looking at what technologies there are and what we need," he explained. "We have a team monitoring the external market place, assessing where we think competitors are going and where we consider we should be positioned. Then we have to translate that into an engineering specification, while considering aspects including complexity as we find the holistic, overall solution."

That accomplished, Pepperell, who has 27 years' experience across JLR, and his team hand over to a program delivery team who are responsible for the detailed engineering of the product: "Which for a brand new engine such as the Ingenium will take three to four years." After that comes the serial development phase.

At Jaguar, another significant element in the development of a new engine is its aural signature, something that must never be subject to palliative action, insisted Pepperell. Much can be done by engineering the exhaust system but he admits that is difficult when developing a small diesel and there is a need for pragmatism: "Frankly, the base diesel will not achieve it."

Could an active electronic sound system provide a solution; it is something that has been discussed within JLR. If it isn't authentic it won't happen, he explained.

Experienced in 132-kW form, the Ingenium diesel, with variable geometry turbocharger (Mitsubishi MHT), does sometimes sound a shade out of place in such a car as the XF, although maximum torque of 430 N·m (317 lb·ft) from 1750 to 2500 rpm helps. It is complemented by a ZF eight-speed auto transmission.

## 700Nm

However, one of the XF's diesel engine options is a V6 diesel. Developed from a Ford/PSAunit, it does make the "right" sound, and with 700 N·m on tap is a very impressive part of the new model's powertrain repertoire. For the new car, the V6's performance has been upped from 202 to 221 kW (271 to 296 hp). It gets the XF to 97 km/h (60 mph) in 5.8 s and achieves a fuel consumption of 5.5 L/100 km and 144 g/km $CO_2$ emissions on the European combined cycle.

Ride and handling solutions that provide very high standards for both have been major selling points for most Jaguars for many decades, and the XF achieves that target far more convincingly than its predecessor.

John Darlington, Vehicle Dynamics Engineering Manager, regards the XF's integral link rear suspension design (developed for the F-Type and also used in the XE) as "our jewel in the crown" with regard to ride/handling.

He also underlines the significance of the XF's "next-generation" damper technology. For city driving, an additional valve permits some fluid to bypass the main valve in the piston, so reducing the damping force and smoothing the ride. As vehicle speed builds away from the city, the valve closes, forcing all of the fluid to pass through the piston to firm the ride.

On V6 versions of XF, a configurable dynamics system (tailored throttle mapping, transmission shift strategy, steering feel, and adaptive dynamics settings) is fitted and controlled via a touchscreen menu.

All of this is part of Jaguar's determined aim to create a car that can carry convincingly the appellation: sports sedan. ■

*Stuart Birch*

# New XF moves to Jaguar's aluminum architecture

**Jaguar XF moves to aluminum body, with a coupe-like look, but with increased rear passenger room, including up to 1.1 in greater headroom.**

**Magnesium is used for XF hood crush beam, as shown, and dashboard cross-car beam.**

Jaguar's best-selling model, the XF, is moving from a steel body to the aluminum-intensive monocoque architecture used on the XE and XJ. It's one of the latest steps the luxury car maker has taken to meet regulatory and emissions demands in Europe and the U.S. The new car, scheduled for introduction in winter, was debuted to the public at the 2015 New York Auto Show. It's a midsize-plus model, fitting between the full-size XJ and the mid-size XE.

The car has a body-in-white that is some 75% aluminum, and much of that aluminum is RC5754, which is primarily scrap recycled with a small percent of new material. The recycled aluminum, which comes from Jaguar's press shops, constitutes much of the 5000 series wrought alloy (aluminum and magnesium) that is used on the XF, where specified to add stiffness (as for underbody reinforcements).

## Steel used for weight distribution

Where steel is used, such as for the rear underbody, in the decklid, and rear-seat-area pan, it serves a number of specific engineering requirements, and it also contributes to a balanced front-rear weight distribution. The end result is a projected 51% front, 49% rear balance.

The steel applications pose an obvious galvanic-corrosion-protection requirement and Jaguar employs a five layer process based on aerospace techniques. First a zinc coating is applied to the steel. Then the joints are made with structural adhesive. Next, the entire body is electrocoated. Following that, seam sealer is applied, and finally the body is painted. There are no body welds, although self-piercing rivets are used.

The bodyside panels are stamped from single sheets of high-strength 6000-series aluminum alloy, which is a wrought alloy that contains magnesium and silicon. A panel weighs less than 6 kg (13 lb).

Jaguar uses specific high-strength aluminum alloys for front and rear crash structures and the A- and B-pillars. The front structure joint with the A-pillar is a large section shaped to distribute crash energy. The B-pillar also gets an

**Rear-seat panel is steel, in part for XF weight distribution, and requires multi-step joint to prevent galvanic corrosion with aluminum body.**

ultra-high-strength (boron) steel reinforcement and is filled with high-density foam. The front suspension towers are high-pressure die-cast aluminum for added stiffness in that area.

Both the dashboard's cross-car beam and the hood crush beam are made of magnesium.

## Precise riveting boosts rigidity

Assembling the side panels to the remainder of the body structure starts with precise locating of the rivets, so the overall result is a more torsionally rigid package. Jaguar claims up to 28% improvement, with the 2016 model specification at 21.7 kN·m (16,000 lb·ft) per degree.

Naturally a Jaguar has to be quiet and the best way is to ensure that is to design noise out. The crash structure behind the front bulkhead, for example, consists of sealed double-wall sections. Foam fillers are placed inside boxed aluminum sections, and when the body goes through the electrocoating ovens, the foam expands to fill. In other locations, sound deadening is a spray-on type that can perform the needed sound reduction with less added weight than mastic sheeting.

The XF is some 80 kg (176 lb) lighter than comparable competition, Jaguar claims, and 190 kg (419 lb) below the previous model with the steel body. Vehicle weight is 3902 lb (1770 kg) for the rear wheel drive model, 4145 lb (1880 kg) for the AWD.

Coefficient of drag was reduced from 0.29 to 0.26. An assist comes from vents in the front bumper that direct airflow smoothly over the front wheels, instead of becoming turbulent around the wheels and creating drag.

The new model is shorter than its predecessor at 195.1 in (4956 mm) long overall, on a 116.6-in (2962-mm) wheelbase. The wheelbase actually is 2.0 in (51 mm) longer, but overall length is 3.3 in (84 mm) shorter. Rear legroom is slightly greater, up 15 mm (0.6 in) to 951 mm (37.4 in). Despite the coupe look, headroom is greater by 1.1 in (28 mm).

There are two 3.0-L supercharged gasoline engines for the U.S. market, both using the same Eaton Roots-type supercharger. One is rated at 340 hp (254 kW) and 332 lb·ft (450 N·m), the other at 380 hp (283 kW) and 339 lb·ft (460 N·m). Both are bolted to the eight-speed automatic, which is required for pairing with all-wheel drive.

A six-speed manual is listed for rear-wheel-drive models and the 340-hp V6, but is not available at launch. A supercharged version of the 5.0-L V8 rated at 550 hp (410 kW) and 502 lb·ft (681 N·m) will be available with a forthcoming Type R performance model.

The electric power steering has been upgraded with improved sensing inputs, better damping steering response to both road irregularities and road camber. The AWD system transfer case is new, with a chain drive for lighter weight and smoother torque transfer vs. the previous gear-drive type.

## Standard, optional suspension variants

The standard shock absorbers contain a second valve to provide velocity- and frequency- dependent damping. During moderate driving conditions, that valve is open and some hydraulic fluid passes through. This additional flow path reduces damping for a smoother ride. At increasing speed, the valve closes and all the fluid has to go through the piston valve, which firms up the ride for better control.

In addition, there's Jaguar's optional active damping system, which senses body movements 100 times/s

**All-wheel-drive transfer case for XF now has chain drive for greater smoothness, replacing gear drive.**

and wheel movement 500 times/s. This electronic system has been improved for better control of damping at all speeds, and incorporates advanced algorithms for road texture detection to provide better traction. Like the F-Type, it can now be driver-configured at the control stack touch-screen.

The XF incorporates a standard advanced traction control system (All Surface Progress Control) that improves drive away on both rear-drive and AWD models in ice and snow. Also included with AWD is torque vectoring by braking, for which the vehicle's electronic stability control modulates torque-on-demand for varying weather conditions.

Integrated with AWD and the active damping system is Jaguar's Adaptive Surface Response. ASR employs algorithms for road-surface-texture detection and uses the information to modify steering response, transmission shifting, throttle response, and stability control operation.

The front suspension is an upgraded double-wishbone design modeled after the F-Type to improve cornering. The rear suspension drops the double-wishbone for a multi-link with an integral link that connects the toe link to the control arm. This separates vertical from lateral forces, so each can be tuned separately, for a better ride.

The XF has a full suite of optional safety features, including automatic emergency braking. For connectivity, it offers the 10.2-in corporate capacitive touch screen with Incontrol Apps for a smartphone. This touch screen and the 12.3-in configurable TFT (thin film transistor) instrument cluster also are available on the Land Rover Range Rover. ■

*Paul Weissler*

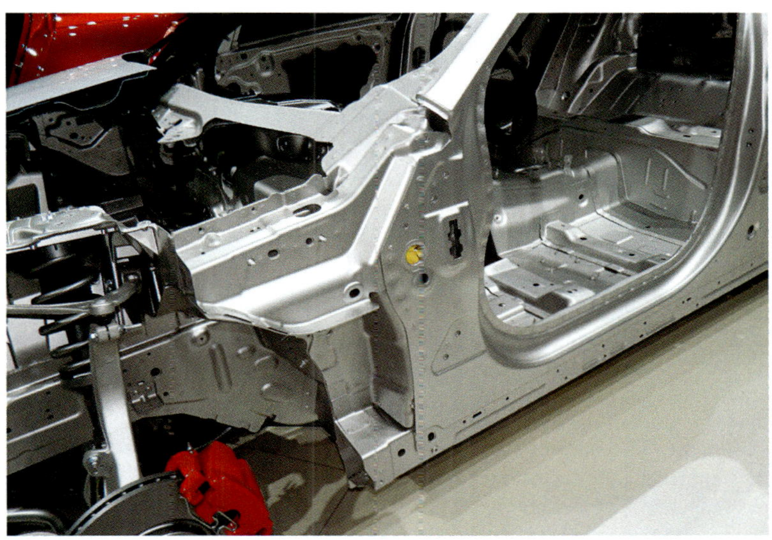

**XF front structure ties into beefed-up A-pillar for optimum distribution of crash energy.**

# Jaguar turns marketing speak into engineering reality for XE

Jaguar's new XE sedan has been created to provide a distinctive marque identity.

"We have a very clear definition of our brand's DNA, and if you drove all our products consecutively, you would be able to experience that through every one," Jaguar Land Rover(JLR) Director of Vehicle Engineering, Mark Stanton, told *Automotive Engineering* at the international launch in Spain of the Jaguar XE.

This may sound like the introduction to a detailed description of how engineering and technology, in parallel with design and styling, set the initial criteria for the car. But it isn't quite like that at JLR; the beginnings of the XE (previously described athttp://articles.sae.org/13509/), arguably Jaguar's most important product to date, can be traced back to the slightly ethereal, emotional, even psychological world of marketing.

"It all starts with marketing words, not technical or engineering terms," said Stanton. "It is to establish a high-level description of the character of the new car, what Jaguar stands for, and initially to do so very much in marketing terms. We take marketing speak – such as 'Grace, Space, and Pace' and 'The Art of Performance'—and turn it into our DNA."

As Director of Vehicle Engineering, Stanton "owned" all the engineering attributes of the XE project and could oversee what was needed to create a product that epitomized those words. "It was quite easy; first we peeled down the 'onion' of attributes of the new model while keeping the word descriptions until we reached the point of physically creating engineering metrics to deliver on those words," he said.

Stanton and the major teams were looking carefully at the XE's potential competitors to ensure that the emerging car would be clearly delineated from everything else in its class: "We set out all the targets, referring constantly to the company 'bible,' which is our DNA."

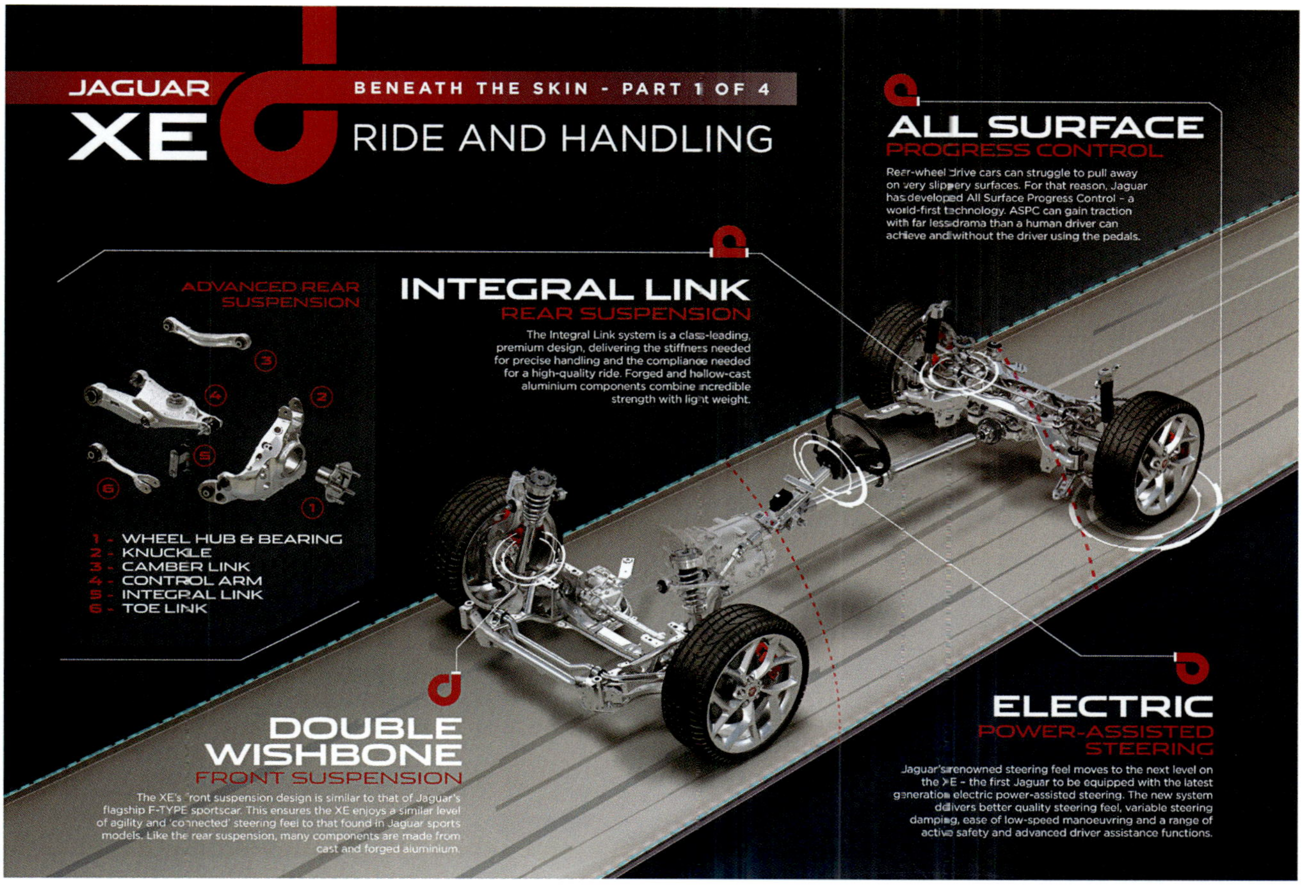


While it is the blend of attributes that achieves what is required, Stanton puts steering at the top of the list: "It is the heart and soul of the XE and the driver's salient connection to the car. Our electric power steering may seem a bit quick to some, but it differentiates us from the competition and through it—within the first 50 meters of driving—you know you are in a Jaguar."

After steering, the engineering focus is on chassis performance as a whole followed by the sound of the engine. That aural requirement is fine for the snarly 250-kW (349-hp) V6—tested on the Navarra racetrack (Circuito de Navarra)—that is shared with the two-seat F-Type, but not so easy with JLR's new 2.0-L Ingenium diesel. Designated AJ200D, it comes with a choice of power outputs: 118 or 132 kW (158 or 177 hp); the former achieves official $CO_2$ emissions of 99 g/km and a best combined fuel consumption of 3.8 L/100 km.

Sampled by this Automotive Engineering editor, the engine provides a smooth almost turbine note at higher revs. "We want to put even more character into our diesels, and we have plans to potentially do that via technology," Stanton added. That usually means active noise systems, but he insisted: "Positive noise: we don't want anything that sounds artificial, it has got to be genuine."

Some of the active sound systems experienced over the past couple of decades have been expensive and complex, he stated: "And they did not deliver that genuine DNA character; our expert owners will know if it's not real!"

Another key criterion of the new XE's creation was weight control, the car using a material mix with a preponderance (75%) of aluminum. Is weight saving plateauing? Stanton is adamant that more can be done: "We are moving on with the possibilities of different grades of material, and technology is further optimizing them. Putting steel in place is no bad thing. It has such good

**As used in the F-Type sports car, the 3.0-L V6 engine fits snugly below the XE's hood and produces the required aural signature.**

**The XE is 75% aluminum but JLR will make more use of carbon in specific areas for future models.**

properties regarding crash performance and NVH—damping noise. Aluminum does not damp noise anywhere near as well. There will be far more use of mixed materials on our models and that will include carbon, which is becoming less costly."

Stanton sees carbon used "locally" in a car's structure for reinforcement in some areas where its characteristics are ideal: "At JLR, we'll see weight continuing to reduce at quite a pace over the next few years."

All versions of the new sedan have plenty of technology, including Jaguar's new InControl infotainment system, which links a Google Android or Apple iOS smartphone to the car. The XE's 8-in touchscreen, which has many modes, is capable of presenting a phone's content. As that phone is updated via apps, so it updates the XE's communications capability to introduce an open platform opportunity. Keeping on-board infotainment systems in step with smartphone advances is seen by Stanton as essential.

"People want, or expect, technology whether they use it or not. They may use only 5% of its capability but would not want the remaining 95% taken away! Making access to that technology for end users is very important, and it must be easy to operate."

Questioned about JLR's medium- to long-term R&D challenges, Stanton explained: "Our expansion is massive; we have 65 programs in place over the next five years. The environment and sustainability top our priority list—and that is in the broadest sense. We're not just talking tailpipe emissions, but everything in which we are involved such as ensuring that our factories and

**JLR's Mark Stanton explained how engineering the XE emerged from "marketing speak."**

**All Surface Progress Control may not be a traditional part of Jaguar's DNA but it is an enhancement for the new XE.**

the disposal of our products meet our environmental requirements. It means total life cycle analysis. But as you start to get into electrification, the equation changes dramatically—such as the construction of batteries and how that is part of vehicle manufacture."

This seems to indicate more JLR hybrids, which Stanton confirmed, but he did not confirm a hybrid XE, partially on the grounds of cost-effectiveness in its targeted price range.

He is very cautious about engine downsizing: "Smaller is not necessarily better; a bigger, lazier engine may be the right solution, under real work load, than a smaller [one]. Our message is all about right-sizing of engines." That is another up-beat opportunity for the marketeers as those 65 programs emerge. ■

*Stuart Birch*

**The XE had to meet the company's DNA specification, with steering quality topping the engineering list.**

# 2016 Optima gets upgraded UVO

**European style has been the theme for the Kia line, and the 2016 Optima continues that approach.**

The 2016 Kia Optima, like previous models, shares chassis and powertrains with theHyundai Sonata, even if some specifics vary according to different product change cycles. However, the company motto remains that "everything you see and touch will be different" between the related models of the two brands. Which means each make gets its very own exterior and interior styling down to switches and knobs, and in some cases, other characteristics such as ride/handling tuning as well.

For the new Optima, featured at the 2015 New York International Auto Show and set to go on sale late in the year, the major technical differences are in the connectivity/infotainment systems. Instead of a built-in modem and a variant of Hyundai's Blue Link, as used on the Sonata, the Kia Optima will be offering an upgraded version of the smartphone-based UVO² (UVO short for "your voice") by adding navigation.

The change is visually obvious, as the 4.3-in control stack screen is replaced by the 8.0-in display that is going virtually across the board on company products.

UVO² incorporates an onboard Microsoft module with Nuance voice recognition. It's on a medium-speed CAN (Controller Area Network) bus but can access the high-speed CAN for trouble codes, as for a vehicle health report.

The navigation system drops Microsoft in favor of one that's Linux-based, and when UVO³ is introduced on a future model, Kia reportedly will swap Microsoft for an open-source Linux to control the entire infotainment system. The navigation is tied to Sirius Travel Link, a subscription service that provides news, sports, weather, local fuel prices, and movie listings via satellite. After a three-month free trial, there's a monthly fee. However, the subscription can be discontinued without losing the navigation system, which is updated with an SD card, via downloads from the UVO website.

Apple CarPlay connects with UVO² via the USB terminal. Google's Android Auto connects with Bluetooth or USB. The full range of Kia eServices apps continues, and Optima now gets geo-fencing, speed alert, curfew alert, and driving score.

The Optima's optional electronics technology package for the SX and SXL also includes a camera array that can produce a 360° (around the car) display on the screen. With this feature already on the Kia list, Hyundai would have to play catch-up when it announces its 2016 Sonata.

As part of the total restyling, the Optima will be the company's first car to offer adaptive headlamps. The electromechanical system pivots the bulb-holding assembly in response to steering inputs, and it is standard with SX and SXL trim packages. Automatic high beam control is included in the optional technology package on those models. ■

*Paul Weissler*

**New 8.0-in center stack touch screen offers navigation, Sirius Travel Link, optional 360° display, and a wide range of smartphone apps. The Infotainment OS includes Apple CarPlay and Android Auto.**

# Most luxurious Range Rover introduced

**The Range Rover SVAutobiography from Land Rover is loaded with luxury touches and features.**

A high-performance vehicle with two-tone paint may sound like something out of the 1960s U.S., an era of some wretched excess. The SVAutobiography featured at the 2015 New York Auto Show by Land Rover is a high-end Range Rover variant (about $200,000), and has a high-output V8. It's produced at the company's Special Vehicle Operations Technical Center in England.

However, it comes across as a quietly, stately model that is being described simply as the most luxurious Range Rover ever. The SVAutobiography supplants the Autobiography Black, and will go on sale in the Fall.

The exterior's top section is the conservative black—and the less ostentatious buyer can select a beltline-down color, such as gray-silver, although there's a choice of nine colors that befits a subdued image.

The interior is made with leather upholstery that includes precision stitching with monogram details and premium wood trim. In addition, there's an optional all-wood sliding panel floor for the cargo area, access of which is by proximity gesture control of the tailgate. The interior electronics include a 12.3-in TFT (thin film transistor) configurable instrument cluster display and a 10.2-in control-stack touch screen with pinch-to-expand.

Under the hood is a supercharged direct-injected 5.0-L V8 rated at 550 hp (410 kW) and 502 lb·ft (681 N·m). This compares with the standard Land Rover supercharged and direct injected V8 rated at 510 hp (380 kW) and 461 lb·ft (625 N·m). The transmission for both is the eight-speed ZF automatic, which features shift changes in just 200 ms, or the choice of manual gear changes with paddle shifters.

Retuning of the Bosch engine management system and recalibrating the bypass valve strategy on the Roots supercharger (increasing boost during high-performance driving) is what enabled Land Rover to increase V8 performance while maintaining current levels of fuel efficiency, the company said. So the U.S. EPA fuel economy should match that of the standard supercharged, direct-injected V8. The 2015 long wheelbase model with that standard supercharged V8 is rated at 14 mpg city, 19 mpg highway.

The long wheelbase, specified for the U.S. model, is 122.8 in (3119 mm), a stretch of 7.8 in (198 mm) from the standard wheelbase. This adds some 7.3 in (185 mm) of rear legroom to the 40 in (1016 mm)

**Rear seats recline on the long-wheelbase SVAutobiography. Table on driver's side is shown deployed from console.**

of the standard-wheelbase Range Rover. Long wheelbase rear seats can recline, and power-deployable tables and a beverage chiller are built into a center console. Overall length is 204.7 in (5199 mm).

The SVAutobiography has adaptive dampers with a response time of 500 adjustments per second, and they can be driver-configured. The air suspension includes a lowering strategy for passenger egress of up to 2.0 in (51 mm).

Like other models in the Land Rover line, the SVAutobiography will have the new InControl Apps smartphone-linked infotainment system using an onboard Intel Quad Core processor. Among the apps are a remote connection to the car, which reads fuel level and remaining vehicle range, tells if the doors are locked or not, and can show the parking location.

**Wood panel slides rearward to ease loading into cargo area.**

The app suite also can remote-start the vehicle and pre-heat or pre-cool the passenger compartment. In the car, the driver connects the smartphone by cable to the console. Optimized smartphone apps then are displayed on the control stack screen with their original look, where they can be touch-actuated by the driver.

Among the many compatible apps are JustDrive, which integrates Twitter, Yelp, and Spotify; Winston (an Apple iOS news reader); MobileDay (which can patch the driver into a teleconference), and Parkopedia, which locates nearby parking, including rates.

Although WiFi and a modem are available with Land Rover products in other countries, a system has not yet been incorporated on U.S. models. ■

*Paul Weissler*

# 2016 Discovery Sport spearheads more efficient Land Rovers

**In development for 3-1/2 years, the compact Land Rover Discovery Sport is already considered a top competitive bogey by other OEMs. For 2016 it switches from Ford power to JLR's new in-house modular gasoline and diesel family.**

The vast, volcanic, and starkly beautiful Highlands interior of Iceland is an ideal place to demonstrate the do-it-all capabilities of Land Rover's all-new Discovery Sport. Two days spent there last January testing early-VIN production vehicles in on- and off-road conditions gave *Automotive Engineering* and other media insight into the work lives of Land Rover engineers. For them, extreme departure angles, tricky split-m surfaces, and icy-river fords are routine development stuff, as proven by how well their versatile new C-segment SUV performs off road.

However, Land Rover's 250-strong engineering team also delivered luxury-car levels of cabin noise and ride comfort on pavement, aimed at the majority of "Disco" Sport customers who will rarely, if ever, venture into the mud or engage low range.

"We've been working closely with our Jaguar colleagues on ride quality and the details of NVH engineering for quite some time now," explained Program Manager Bob Prew. He said development of the new vehicle (known internally by its L550 program code) targeted two pairs of competitors: Audi Q5 and BMW X3 were the "premium" bogies, and Honda's CR-V and the Hyundai Sorento were the "non-premiums"—the latter receiving Prew's praise for their overall five-seater's execution.

## New multilink rear

The newest Land Rover replaces the LR2/Freelander and uses a modified iteration of the Evoque's LR-MS unibody architecture that Land Rover engineers said will underpin subsequent models. The compact SUV is 181.1 in (4599 mm) long and 81.5 in (2069 mm) wide and rides on a 107.9-in (2741-mm) wheelbase—3.25 in (82.5 mm) longer than Evoque's and 10 in (254 mm) longer overall. The taller greenhouse benefits passenger ingress and interior room, which feels as spacious as any vehicle in the class.

Two seating configurations, five- and seven-seat (called "five plus two") are offered. The tight third row, which folds flat, sits on a "stadium" perch about an inch (25 mm) above the second row and is only practical for small kids and pets. With both seats fully stowed, the vehicle offers 60 ft3 (1700 L) of cargo volume, claimed to be class-leading.

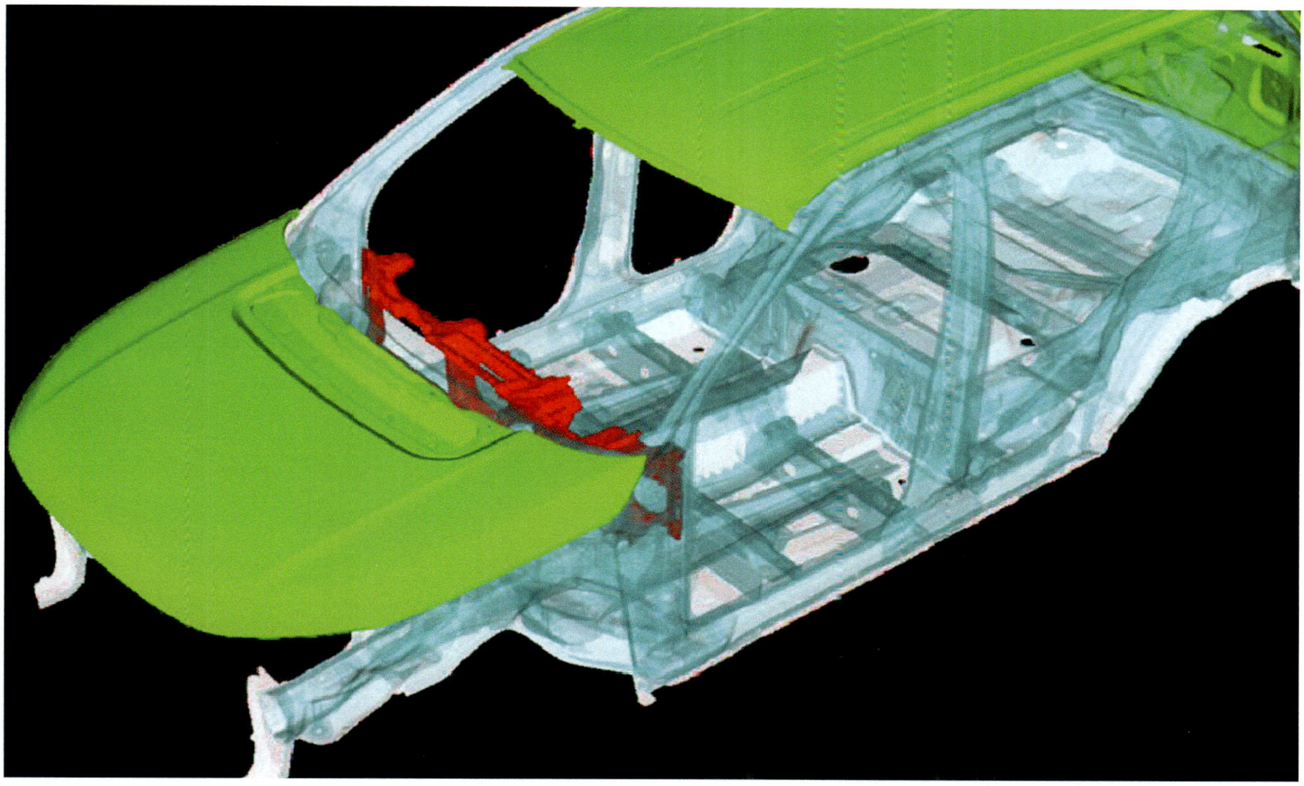

**Body-in-white features 20% HSS boron alloys, with magnesium cross-car structure and aluminum hood, roof panel, front fenders, and liftgate.**

The Discovery Sport's steel-intensive body structure features 20% high-strength boron alloys, with aluminum hood, roof, front fenders, and liftgate, and a magnesium cross-car structure behind the cowl. Base curb weight is 3854 lb (1748 kg). The hood assembly includes an airbag module mounted on the windshield side for meeting ped-pro (pedestrian impact) requirements. The vehicle earned a five-star safety rating in Euro NCAP testing.

Prew's team extensively modified the Evoque's structure to package a new compact multi-link rear suspension—itself designed to enable the third-row seat to be folded down into the cargo well with enough room to also accommodate the space-saver spare tire (on five-seat versions; the spare is underneath the vehicle on 5+2 models). "It was critical that we keep the rear load floor at a practical height," he said. The modular MacPherson front suspension carried over from Evoque uses hydraulic rebound stops on the struts. Its upper and lower control arms are in aluminum and steel, respectively.

Variable-ratio electric power-assist steering is fitted, which Prew said enables an automatic parking assist function. The Disco Sport's steering requires 2.43 turns lock-to-lock, with a 38.9-ft (11.9-m) turning circle. Foundation braking is by four vented discs, 12.8-in (325-mm) front and 12-in (300-mm) rear.

## New Ingenium engines power 2016 models

The Discovery Sport is available in both two- and all-wheel-drive configurations. For 2016 the new Jaguar Land Rover "Ingenium" 2.0-L family of modular turbocharged gasoline and diesel engines (http://articles.sae.org/13353/) replaces the previous Ford-sourced EcoBoost and Duratorq ranges used on MY2015 vehicles. Developed by JLR in-house, the Ingenium range features a common 0.5-L cylinder displacement, rolling-element camshaft bearings for reduced friction, and stop-start/mild-hybrid functionality.

North American market vehicles are expected to use the gasoline engine initially, paired with the incumbent ZF 9HP nine-speed automatic transmission. A Getrag M66EH50 six-speed manual is available in global markets.

There are two different AWD setups, neither equipped with a transfer case. Standard is a BorgWarner-sourced Haldex system; optional is the Land Rover Active Driveline, an axle-disconnect system co-developed with and supplied by GKN Driveline. Active Driveline is claimed to reduce driveline parasitic losses by 75% compared with the Haldex system. The GKN system also features torque vectoring-by-braking, controlled via the DSC (stability control) module at rates up to 100 times/s.

To deliver a low-range 4x4 function, the 9HP is fitted with a super-low "crawl" first gear ratio available only in off-road conditions. For typical road use the vehicle starts in second gear. Locking differentials are not offered on the new Disco Sport.

Land Rover's patented Terrain Response 2 offers drivers several automatic operating modes for tackling various road conditions and traction situations, with the option of manual override. The technology (which includes a new pitch-rate sensor and upgraded Hill Descent Control that is claimed to be three times faster than that used on the LR2) has settings for General, Grass/Gravel/

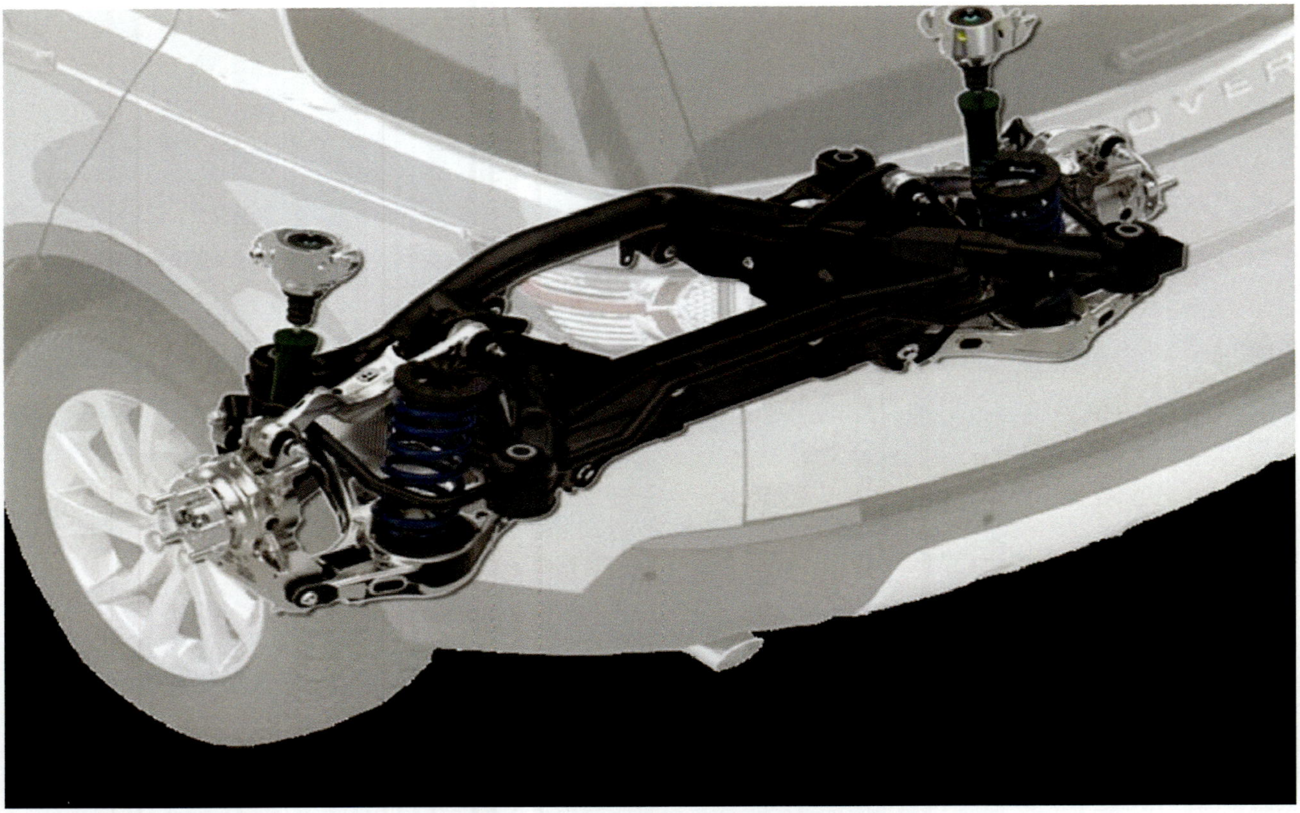

**The design of the Disco Sport's new rear suspension module allows 320-mm (12.6-in) axle articulation.**

Snow, Mud/Ruts, and Sand vehicle use. The system's algorithmic maps configure throttle, steering response, transmission gear ratio, center diff, and the braking/DSC for optimum performance.

Terrain Response Control 2 is not offered on the front-wheel-drive version, which will be JLR's fuel-economy play for this segment.

## 600-mm water fording

Despite its on-road design bias, the 8.3 in (212 mm) of ground clearance and up to 24 in (600 mm) of water-fording capability—the latter an improvement of 3.9 in (100 mm) over the LR2—gives the new Disco Sport more than respectable off-road capability, as our Iceland cold-weather testing confirmed. Approach and ramp angles, 25° and 21°, respectively, are slightly less than those of the LR2, while the 31° angle-of-departure carries over. But for owners who crave more rock-crawling capability, a section of the front bumper can be removed to increase the approach angle.

The Discovery Sport's interior design and trim are straightforward and fairly plain. Potential distractions appear to be minimized. An 8-in touchscreen that is shared with the new Jaguar XE sedan dominates the center of the IP and controls vital infotainment and navigation functions for the vehicle's three trim packages. The SE base model includes rear parking sensors, rearview camera, and 18-in wheels among many other features. Next up is the HSE, which adds 19-in wheels, a panoramic glass roof, and upgraded audio and interior trim.

Top-of-range HSE Lux models can be had with a Vision Assist package that includes adaptive xenon headlights, automatic high beam control, a surround-view camera system, blind spot monitoring, rear cross-traffic alert, and parking assist. ■

*Lindsay Brooke*

# See-through A-pillars, ghost navigation, and Discovery Sport extreme testing

**Land Rover's Discovery Sport was tested in some of the world's toughest environments. It is available with a 140-kW/420-N·m bi-turbo 2.2-L diesel engine, but new power units are planned, and the company has an expanding hybrid program. The car has a Cd of 0.36.**

When Land Rover wants to reveal its latest engineering and technology, it invariably does so in extreme driving conditions; conditions that reflect its rigorous test programs.

For the new all-wheel-drive Discovery Sport (see http://articles.sae.org/13602/), it chose Iceland, unpaved mountain roads, and (although it claimed no control over the environment), a 100-km/h (62-mph) whiteout blizzard with minimal visibility, the rear lights of the car in a "buddy-buddy" convoy just a few meters ahead often totally obscured and at times impossible to follow. With the urgent need on tight corners to peer intensively around the A-pillars, any aid to visibility would have been welcome as the car's tires, with a total of 800 studs, struggled for grip.

So it was not wholly a coincidence that as the testing Icelandic challenge progressed, Jaguar Land Rover (JLR) revealed that it is now focused on a radical visibility enhancement research program that, if successful, will create what it calls "Follow-me Ghost Navigation" projecting a head-up virtual image of a car ahead for the driver to follow, and even more radically, "see-through" A-pillars using the interior of the pillars as screens to present a real-time video of the normally blanked scene.

"See-through" A-, B-, and C-pillars and a "Follow-me Ghost Navigation" system are being developed by JLR.

The Discovery Sport was tested in some of the world's toughest environments. It is available with a 140-kW/420-N·m bi-turbo 2.2-L diesel engine, but new power units are planned, and the company has an expanding hybrid program. The car has a Cd of 0.36.

Instead of trying to follow the taillights of a real car (pictured) as a guide to a destination, JLR has come up with a "ghost" image that will always be in sight just a few meters ahead.

Dr. Wolfgang Epple, JLR Director of Research and Technology, wants to see, quite literally, 360° visibility in all the company's vehicles, with the A-pillar system extrapolated to B- and C-pillars: "If we can keep the drivers' eyes on the road ahead and present information in a non-distracting way, we can help them make better decisions in the most demanding and congested driving environments."

He explained that a screen would be embedded in the surface of each pillar inside the car, taking a live video feed from cameras covering every angle outside the vehicle to compensate for normal pillar blind spots.

The system would also be linked to driver head movements or use of the turn indicator to trigger pillar images. Other solutions to A-pillar blocking have been considered by automakers, including Volvo's open girder concept.

In effect, the JLR system would create a 360° virtual windshield and would also have a link to the Cloud, enabling it in an urban area to present, in HUD form, information such as parking space availability or gas station prices.

Because they are camera-based, the systems involved might be struggling in the Arctic conditions in Iceland, which had the windshield wipers building up ice

JLR Research and Technology Director, Dr. Wolfgang Epple, wants the company's vehicles to provide 360° cabin vision.

Discovery Sport Program Director Paul Cleaver says there is "lots of engineering complexity" in the rear section of the car, including integral multilink suspension, to meet seven-seat versatile packaging needs.

layers despite the permanent use of windshield heating. Heated wiper blades might be another R&D path to consider.

The decision to use Iceland, with its highly unpredictable weather patterns (temperate through near-Arctic) in mid-winter, was unusual for any motor manufacturer, although Volkswagen and Mercedes-Benz have used it in more clement seasons.

But it is the sort of extreme weather experienced that is the stuff of SUV testing, and Iceland delivered a soupçon of this to *Automotive Engineering* editors who were invited by the company to drive production examples of the new seven-seat Discovery Sport.

Although "testing" is the word that links every aspect of vehicle creation and development, it may also be the least appreciated. It is very easy for a manufacturer to reel off thermal, topographical, distance, and a host of other facts and figures about the rigors—virtual and real—to which their latest products have been subjected, but it can be difficult to realize exactly what all this means. The very tough Iceland experience achieved that.

The Discovery Sport is based extensively on Evoque architecture, but with some significant differences, notably the use of fully independent integral multi-link rear suspension, a necessary configuration to facilitate the fitment and stowage of a third pair of seats that can fold into the floor.

Packaging was a major criterion of the design, says Program Director Paul Cleaver: "Overall length was really important; we wanted to create a compact SUV at sub-4.6 m (the production car measures 4590 mm). Everything else in the segment with an extra row of seats was over that length. The decision was pivotal in the way we engineered the back end. The Discovery Sport had to be a wheel-at-each-corner, short overhangs, and compact stance solution."

The compact multi-link rear axle was essential. As well as its dimensions, its pluses include on-road agility, claimed class-leading wheel travel and axle articulation at 340 mm (13.4 in), and required levels of on and off-road composure. Its use contributes to reduced road noise entering the cabin. Because the suspension turrets make only a minimal intrusion into the luggage area, they allow row-two seating to slide and recline past the turrets for maximum leg-room.

Squeezing seven seats (row three folding flat) was a main criterion for the Discovery Sport.

**Interior of the Discovery Sport.**

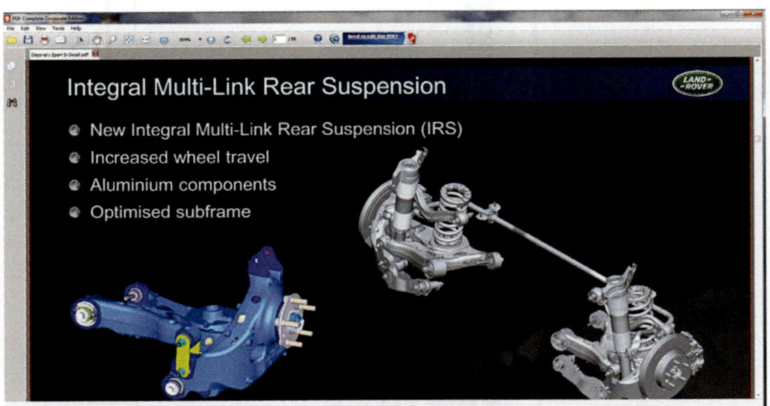


**The use of integral multi-link rear suspension facilitated improved packaging for the Discovery Sport.**

The lower control arm and rear suspension knuckle are produced from thin-walled hollow aluminum castings. The rear suspension is mounted on a steel subframe, which is light and stiff to promote improved steering response.

Another significant technology aspect of the Discovery Sport is the incorporation of a pedestrian airbag triggered by pressure-tube sensors mounted between bumper cover and beam, which is designed to detect impacts with adult pedestrians. Should any impacts be detected between 24 and 48 km/h (15 and 30 mph), an airbag concealed by a small scuttle panel at the base of the windscreen deploys in no more than 60 ms, drastically reducing the risk of serious pedestrian injury, explained Cleaver.

The scuttle cover is attached to the airbag using two tethers and forms part of the airbag system in helping to spread the load evenly across the airbag, avoiding obscuration of the driver's vision during deployment and reducing airbag volume to 110 L (3.9 ft$^3$). Rigorous testing ensured that it would not deploy in extreme off-road situations, added Cleaver: "The innovative pedestrian-protection system is not only unique in the SUV class, it's also yet to debut in most other segments." The new Land Rover has achieved a 5-star EuroNCAP safety rating.

The Discovery sport comes with a choice of six-speed manual or nine-speed ZF automatic transmission. It has very comprehensive electronic chassis support systems and uses a Haldex Generation 5 system, but does not have a low ratio transfer box.

Would that have helped in the exceptionally challenging conditions in Iceland experienced by *Automotive Engineering*? Possibly. But Land Rover's support teams were on hand with Defenders to pull any errant vehicles back into line, another—albeit low priority—aspect of vehicle testing: ensuring that the media survive. ■

*Stuart Birch*

# Discovery Sport Introduces Evoque-based Land Rover

**Discovery Sport is based on Range Rover Evoque powertrains and front end, with a new rear including a multilink rear axle.**

Weight reduction has been an ongoing theme for Land Rover in recent years, with significant reductions for both the Land Rover Discovery and Range Rover. With the launch of the Discovery Sport, unveiled at the 2014 Paris Motor Show, Land Rover is extending the weight reduction theme with engine downsizing.

It's the first Discovery badged vehicle in a decade not to be offered with a V6 or V8 engine. In fact, the Discovery Sport has much more in common with the Range Rover Evoque, with which it shares powertrain options and many structural components.

Although the car has been launched with the 2.2-liter four cylinder diesel, which can trace its roots back to the company's days under Ford ownership, it's certain that replacement engines are on the horizon, now that production of a new family of Jaguar Land Rover turbocharged 2.0-L four-cylinder Ingenium gasoline and diesel engines is due to begin in early 2015. For now the 2.2-L diesel is offered with power outputs of 150 hp (112 kW) for the eD4 and 190 hp (142 kW) for the sD4. A 2.0-L gasoline engine is also available rated at 240 hp (179 kW).

The Discovery Sport structure is based on a steel monocoque, but with a hood, roof, and tailgate formed from aluminum. A single-piece cross-vehicle magnesium beam is used, saving around 7 kg (15 lb) compared with a fabricated steel alternative. Curb weight starts from 1765 kg (3890 lb).

A new lightweight structure is just one of the technical highlights. It's the first smaller Land Rover model to offer up to seven seats, or five plus two (5+2) seating as Land Rover describes it. To accommodate this, Land Rover has designed a new multilink rear axle.

The company claims a first in class for the hood-mounted pedestrian airbag, fitted as regular equipment, along with Autonomous Emergency Braking (AEB). Although AEB is not yet mandatory for passenger cars in Europe, it will be impossible for new models to gain a five-star Euro NCAP rating in the future if the system is not fitted.

The AEB system uses a pair of stereoscopic cameras to sense objects ahead of the vehicle up to 80 m (262 ft) ahead. The cameras are mounted near the rear-view mirror behind the windshield. "When it monitors a potential collision risk, it operates between 5 and 30 km/h," said Cleaver. The system will give an audible warning and a visual warning in

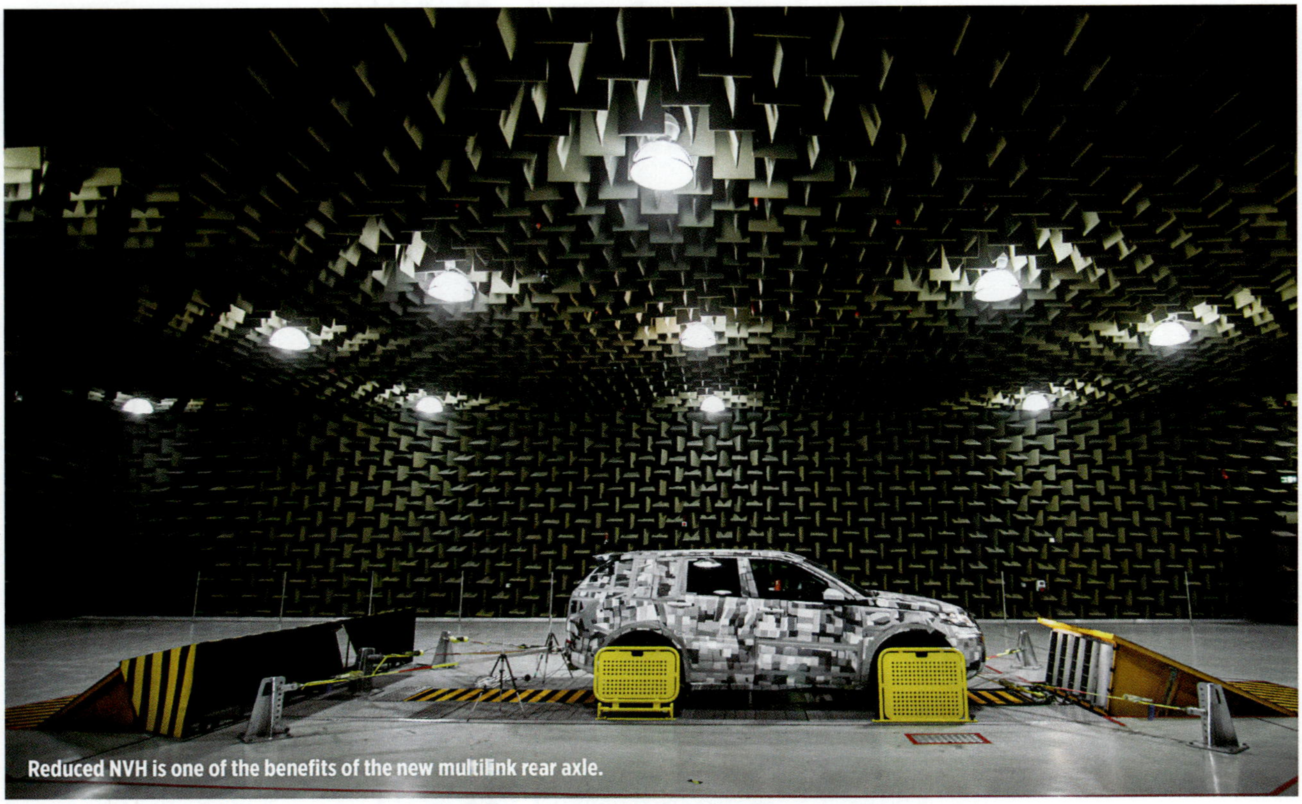

Reduced NVH is one of the benefits of the new multilink rear axle.

the instrument cluster or head-up display. "If the driver doesn't react, it precharges the brakes; if the driver still doesn't react, it automatically brakes the vehicle. Up to 35 km/h (21.7 mph), it will avoid a collision altogether and will mitigate collisions up to 80 km/h (49.7 mph)."

Four-wheel drive, using an electronically controlled Haldex center coupling, is the regular offering at launch, but this will be supplemented later in 2015 with a two-wheel-drive diesel-powered model badged eD4, with projected CO2 emissions of 119 g/km, measured on the New European Driving Cycle (NEDC).

Land Rover claims an SUV first with the hood-mounted Pedestrian airbag

The Haldex coupling will automatically decouple the rear differential in regular on-road driving, reducing frictional losses in the driveline. When other drive modes are selected using the Terrain Response system, the center coupling is automatically engaged. Four-wheel-drive Discovery Sports will have the choice of a ZF nine-speed automatic transmission or Getragsix-speed manual. Gasoline powered models are available with the ZF nine-speed automatic transmission.

"The main objective for us has been to develop a very compact SUV," Paul Cleaver, Discovery Sport Vehicle Program Director, told *Automotive Engineering*. "A key priority was not to exceed 4.6 m (15 ft) in overall length. When you look at how this car compares with other premium SUVs, it's actually 39 mm (1.5 in) shorter than the Audi Q5, for example. If we look in the rear of the car, this is a 5+2 seater, with rear seats designed for teenagers and occasional use for adults.

"In terms of engineering, Discovery Sport is based off the Evoque platform. So the powertrain architecture, front end, transmission, driveline, all the technologies are based on Evoque, but everything from the heel board backwards is new, including the rear suspension system.

"We've got a new integral link rear suspension, which we have designed from scratch. The concept is the same as we use on our bigger products, the Range Rover and Range Rover Sport. We've selected

**Hood, roof and tailgate are manufactured in aluminum to aid weight reduction**

that system to afford us the space that we needed in the rear of the car to package the seats. So all the suspension is new and given that we've designed it from scratch, we've optimized NVH (noise, vibration and harshness) and dynamics.

"We've taken the opportunity to optimize those components from a weight perspective. The knuckle and the lower control arms front and rear on the suspension system are hollow cast aluminum components to take weight out of the vehicle.

"So a lot of the focus on the architecture has been around the back end of the car, firstly to get that very compact stance. Proportionally we wanted the car to be the way it is from a design perspective, but then if you look at the occupant package that we've managed to deliver with the +2 seats and the slide-and-recline second row seats, we didn't want to give anything away in terms of interior occupant space."

The second row seats can slide fore and aft through 160 mm (6.3 in). With the second-row seats set in the rearmost position, occupants have 1011 mm (39.8 in) of knee room, compared with 1025 mm (40.4 in) for the Range Rover Vogue. The wheelbase is 81 mm (3.2 in) longer than the Range Rover Evoque, while the car is 91 mm (3.6 in) longer overall than the Land Rover Freelander.

Land Rover claims that the Discovery Sport is the first SUV on the market to offer an external pedestrian airbag. The airbag has a volume of 110 L (3.9 ft$^3$) and operates at speeds between 25 and 50 km/h (16 and 31 mph). Pressure sensors are located in the front fender, connected via a pressure tube. "When the system detects enough force on the pressure tube, the airbag is deployed within 50-60 milliseconds," said Cleaver. "The bag is deployed across the trailing edge of the bonnet and up and around the A-pillars." ■

*John Kendall*

SAE Eye on Engineering Quick Take:

## New Land Rover Discovery Sport

Watch the video at http://youtu.be/ccZh9orCgC8

Not everyone needs a Land Rover. But when driving conditions are tricky, it's good to be in Britain's famous sport utility vehicle. Senior Editor Lindsay Brooke recently spent two days in Land Rover's new Discovery Sport in Iceland. SAE Eye on Engineering can be viewed at http://youtu.be/ccZh9orCgC8. It also airs in audio-only form Monday mornings on WJR 760 AM Detroit's Paul W. Smith Show.

Access archived episodes of SAE Eye on Engineering at www.sae.org/magazines/podcasts.

# Lexus sharpens best-selling RX for 2016

**New Lexus RX has sharply-chiseled look across front and sides. Crossover was introduced by Lexus executives at 2015 New York Auto Show.**

The Lexus RX luxury crossover has been the make's best seller, accounting for 25-30% of sales, and the 2016 model introduced at the 2015 New York Auto Show displays a new styling treatment that features crisp lines for a chiseled look. There's now an F-Sport model for both the all-wheel-drive and hybrid models; the F-Sport previously had been available only for the non-hybrid. The crossovers go on sale at the end of the year.

The new edition, a fourth generation in the mid-size vehicle's 17-year history, is 4.7 in (119 mm) longer at 192.5 in (4890 mm) overall, on a 2.0 in (51 mm) longer 109.8-in (2789-mm) wheelbase. The C-pillars are partially blacked out for the appearance of a "floating" roofline.

The 3.5-L direct-injected V6 has been reworked to increase performance to 300 hp (224 kW) from the previous 270 hp (201 kW), and continues to be bolted to the eight-speed automatic transmission. The RX 450h hybrid has an Atkinson-cycle version of the 3.5-L V6, rated at 245 hp (183 kW), but with the electric motor, total output will be rated at almost 300 hp (224 kW). Both engines will operate on regular gasoline, and Lexus is promising improved fuel economy for the gas-engine-only model.

## Safety systems pricing

Toyota announced an aggressive pricing strategy for its safety systems packages, with the one for Lexus to list for under $650.

Lexus Safety System-plus (LSS+) consists of a pre-collision system that uses millimeter-wave radar and a camera to identify pedestrians as well as vehicles and activate a brake assist system. If the driver doesn't respond to warnings by hitting the brakes, automatic emergency braking is initiated.

It operates at speeds of 7-50 mph (11-80 km/h) with the pedestrian detection strategy and can reduce vehicle speed by about 19 mph (31 km/h). The reduction, based on testing vs. a vehicle going at 19 mph and a stationary vehicle/pedestrian, obviously is subject to such variables as road/weather. The vehicle detection comes into play at any road speed from 7 mph and above, and reduces road speed by about 25 mph (40 km/h).

**F Sport console knob to rear of cupholders provides ride-control adjustments including Sport+ for active anti-sway system.**

In addition, LSS+ incorporates lane-departure alert and, using the front camera of a four-camera 360-degree viewing group, provides automatic high beam control. Furthermore, using the front camera and radar, the RX provides adaptive cruise control.

The control stack incorporates Lexus' 12.3-in screen that, in addition to navigation, comfort features, and infotainment, can display a 360-degree view around the vehicle. The interior features a TFT (thin film transistor) multi-information instrument cluster display.

LED taillamps wrap around the sides at the rear, and LEDs also take over front lighting. Fog and daytime running lights are LED, and triple-L-shaped LEDs are available for headlamps.

A proximity sensor is built into the tailgate. With the remote key in the motorist's possession, swiping a hand close to the tailgate's Lexus emblem and pausing briefly will unlock the tailgate and allow it to open.

## Michelin tires wear pattern

The RX will be using Michelin's new Premier tires for 18- and 20-in wheels. The tires have a high-traction rubber that as the tread wears is engineered to expose hidden rain grooves and then cause them to expand. So the wet grip and overall tire safety are maintained throughout much of the tread life. As a styling touch, body-color-coded inserts are available for the wheels. The F Sport has a specific trim package including unique grille, 20-in alloy wheels with a graphite finish, and black side mirrors.

An active anti-sway system, which produces flatter cornering, is borrowed from Lexus' GS and some hybrids. The system incorporates sensors signaling an electronic control unit to operate motors working through reduction gears to torque the anti-sway bars to minimize body roll. The system is enabled by a center console control knob with Eco, Normal, Sport, and Sport S+ positions. ■

*Paul Weissler*

# High-performance Otto-to-Atkinson V8 edition of Lexus GS introduced at NAIAS

**The GS-F gets a more aggressive front appearance to promote its sports sedan image. Note orange caliper projecting through front wheel.**

Lexus continued on its path to a high-tech performance car image from its long-held soft-luxury reputation, with the introduction at the North American International Auto Show (NAIAS) of the newest sports sedan in its F line, the 2016 GS-F.

The F designation is Lexus' mark of race-tinged performance, and the new edition in the GS series comes on the heels of the RC-F sport coupe, with which it shares an advanced V8 engine and torque-vectoring rear differential. Toyota is serious enough about the performance image to tell the world of motorsports that it will be going full-throttle into the FIA's GT3 racing, officially named "Group 3: Cup Grand Touring Cars."

The GS-F is, like other models aspiring to compete in the German-dominated sports sedan market, a product of suspension development at the Nurburgring, but also on Japan's Fuji Speedway. It rides on wide 19-in forged aluminum wheels with 15-in (381-mm) front rotors and attention-getting Brembo six-piston bright orange calipers. The custom-tuned suspension includes underhood hand-turn adjusters for the shock absorbers.

And because it's a Lexus, the five-passenger sedan production model also will have all the safety features of the more sedate models, including Pre-Collision with radar-guided automatic braking and pedestrian detection. But it's the engine and drivetrain technology (which actually debuted in the RC-F immediately before NAIAS) that draws the engineering attention.

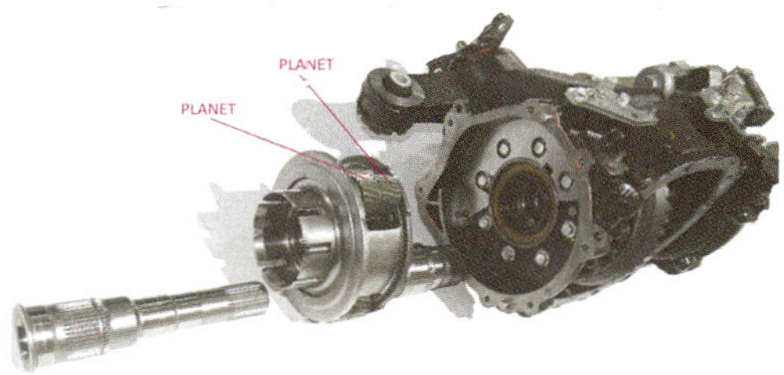


In addition to a clutch pack, at each side of the torque vectoring differential is a planetary gearset with two sets of planets, each having a slightly different ratio.

Underhood look includes painted parts for a performance image. A snap-on plastic panel has been taken off to get a look. It pops off easily for the "show business" effect.

New electric intake valvetrain phaser provides smooth transitions between Otto cycle and Atkinson cycle operation.

## Otto to Atkinson cycling

The 5.0-L V8 is rated at 467 hp (348 kW), a hearty 51 hp (38 kW) more than the previous generation. Peak torque is 389 lb·ft (527 N·m), and the engine carries a 7300-rpm redline. It's bolted to a quick-shifting version of the eight-speed Lexus automatic. Much of the performance increase is from cylinder head porting and valvetrain modifications.

Although this performance boost might be expected to impact fuel economy, Lexus is saying it won't. This is thanks to a major change in engine calibration plus use of a new intake valvetrain electric cam phaser that permit the engine to switch from Otto cycle to Atkinson cycle operation when engine load is low. The engine itself is the latest version of the V8 that was developed for the original Lexus performance car, the IS-F.

The LS460 has had an electric intake phaser for some time. But the new one is a further development for the seamless transitions between Otto cycle and Atkinson cycle (late intake valve closing to simulate a longer power stroke, for improved efficiency). That will enable the GS-F to post EPA window sticker numbers comparable to the standard GS V8, Lexus said.

The direct-injection fuel system is a comparatively high-pressure design, with peak pressure at 5000 psi (345 bar).

## Torque-vectoring sans ABS

The torque-vectoring rear differential (TVD) is a step beyond the type that can add power to a wheel to help power a car coming out of a turn. In other systems, the torque-vectoring control also uses the anti-lock brakes/traction control actuators (ABS/TC) to brake the inside wheel (reducing torque) to guide the car when it is entering a turn.

The Lexus differential has a multi-disc clutch pack at each wheel, controlled by a separate actuator with ball-ramp device to vary the application pressure. Each pack regulates torque flow through a planetary gearset with two sets of planets, each planet with a different ratio to bias the system. A ratio change occurs when a clutch pack is actuated.

On command from the skid control module (VDIM—Vehicle Dynamics Integration Management module), the system can add or reduce torque to each wheel, so there is no power loss from use of the ABS/TC operation at any time. Entering a turn, torque is reduced evenly to both rear (drive) wheels for deceleration stability. As the car approaches the apex of the turn, torque is further reduced to the inside wheel to help the car rotate slightly. Closer to and at the apex, the system transitions to positive torque, with somewhat more going to the outside wheel to promote stability. As the car starts to accelerate out of the turn, the torque to the outside wheel is increased further.

When the system operates to guide the car through a turn, it effectively is performing rear-wheel steer.

The TVD can be pre-selected for any of three modes: 1) the default standard, intended to provide a balance for everyday (albeit exuberant) driving; 2) slalom, which increases the responses for more nimble steering and vehicle agility; 3) track, a tuning for a high-speed circuit. ■

*Paul Weissler*

# Mazda enters subcompact crossover segment with 2016 CX-3

Mazda's designers pushed the front fenders' peak rearward to the base of the A-pillar to convey a sense of extended body size, so the CX-3 looks bigger and more expensive than it is.

Mazda dove into the burgeoning market for subcompact crossover SUVs with the 2014 Los Angeles Auto Show introduction of the CX-3, a stylish new tall wagon built on the platform of the new Mazda2/Demio.

As with other recent Mazda introductions, the CX-3 features the company's Skyactiv drivetrain and platform technology, an all-encompassing philosophy of coordinated optimization of efficiency that provides useful benefits from the aggregation of tiny improvements.

Mazda describes the CX-3's chassis as a simplification of the Skyactiv architecture, one that employs increased use of high-tensile steel and optimized cross-section shapes to provide light weight and positive damping characteristics.

While there is no current CX-3 for comparison, the new Mazda2's body-in-white is 7% lighter than that of the old car, while it boasts a 22% improvement in torsional rigidity, providing some perspective for the CX-3's derivative chassis.

High-strength steel accounts for 63% of the CX-3's unibody, with 29% 780-MPa (113-ksi ultra-high-strength steel, and 4% 1180-MPa (171-ksi) steel for maximum resistance to intrusion in a crash.

As with the Mazda2, the CX-3 uses MacPherson struts in the front and a simple torsion beam for the rear suspension. No details are yet available, but the company promises that fine-tuning of the details yields improved control and response when compared to the previous-generation Mazda2 platform.

Though the CX-3 and Mazda2 share the same 101.2-in (2570-mm) wheelbase, at 168.3 in (4275 mm), the CX-3 is 8.5 in (216 mm) longer. It is also 1.8 in (45 mm) wider and stands 2.1 in (53 mm) taller.

The company says that it refined the CX-3 cabin noise by controlling suspension resonance, tuning the shape and hardness of engine mounts and adjusting their positions, and locating sound insulation and deadening materials in the optimal locations.

Special attention was given to the all-wheel-drive models, which see reinforcement to the crossmember connecting the rear subframe and strengthening the spare tire pan with optimized shaping of the beading. Computational fluid dynamics work is credited with reducing both aerodynamic wind noise and fuel-economy-sapping drag.

The CX-3 is available in both front- and all-wheel-drive configurations. The company claims a world-first with the inclusion of a front wheel slip warning detection system that it says monitors driving conditions and tries to predict drivers' intentions to keep the CX-3 on course. The all-wheel-drive system uses the active torque control coupling seen on the CX-5, mated to a new power take-off unit on the front transaxle and a new, more compact and lightweight rear differential.

**The dark red accents of the door trim armrests, along with the stitched finish of the soft cover on the instrument binnacle, give the CX-3's cabin a rich, detailed appearance.**

All U.S.-market cars are powered by a 2.0-L I4 from Mazda's Skyactiv engine family. As installed in the 2015 Mazda3, that engine is rated at 155 hp (116 kW) and 150 lb·ft (203 N·m). All U.S.-spec CX-3s will be fitted with the company's six-speed automatic transmission, which uses a small torque converter for launch from a standstill and a computer-controlled clutch at higher speeds.

Global market CX-3s will likely see the availability of the 1.5-L gasoline and diesel engines, as well as a six-speed manual transmission, that are offered in the Mazda2 in those countries.

Inside, the CX-3 features the infotainment interface introduced with the 2014 Mazda6, the flip-up dashboard-mounted head-up display seen on the Mazda3, along with all of the safety-related technologies available on those cars, most notably, the forward-collision-warning radar.

The company points to the high quality of cabin materials as an advance in the class, and inspection on the show floor confirmed that upholstery material quality, the coordinated piping and the stitching convey a rich, expensive appearance that is unexpected in the subcompact segment.

Additionally, though the CX-3 is small, the rear seats proved useable by normal adult males, though ingress and egress through the small door openings is tight.

Mazda says the CX-3 goes on sale first in Japan, starting in spring 2015, arriving in global markets later in the year. ■

*Dan Carney*

**The U.S.-market CX-3 will use the same 2.0-L Skyactiv I4 seen previously on the Mazda2.**

# 2016 Mazda MX-5 stays true to its roots

**MX-5's development team attacked every gram of potentially excess mass to create a vehicle that is more compact and 150-lb lighter than its not-exactly overweight predecessor. The body exterior and structure are a combination of high-tensile (780 MPa or higher) steel and aluminum alloy. Competitors take note: the seats are significantly lighter than on the outgoing Miata.**

In the last 26 years, various "affordable" small sports cars have come and gone while the Mazda MX-5 Miata remains the unwavering benchmark of the genre. It's remarkable how the company's designers and engineers kept the model so true to its purpose—a driver's car in the spirit of the first Lotus Elan—throughout its evolution from the 1989 original (known internally as the NA), through its NB (1997), and NC (2005) successors. Hardcore Miata fans may pick nits about one generation versus another, but the MX-5 has lost none of its mojo atop the purist roadster podium.

Expected to soon reach the 1-million-units production milestone, the MX-5 has entered its fourth generation as a 2016 model. The all-new ND's development cycle was protracted because getting it right was so essential—MX-5 is to Mazda what 3 Series is to BMW, and what Wrangler is to Jeep. Any major misstep would be catastrophic to the brand, as Program Manager Nobuhiro Yamamoto's team clearly understood. Their mission: Take the holistic goodness of the previous-generation car—mechanical simplicity, 50/50 front/rear weight balance, lively vehicle mass-to-power ratio, communicative and nimble chassis dynamics—and improve on it while keeping cost in check.

To the delight of sports-car lovers—and the FCA team charged with developing the upcoming Fiat 124 Spider that's derived from the MX-5—the ND team hit its bogies and then some. Most encouraging is that Mazda engineers proved that a new model doesn't have to be larger and heavier than its predecessor. The 2016 Miata is 105 mm (4.1 in) shorter, 20 mm (0.8 in) lower, and 10 mm (0.4 in) wider overall than the '15 car, while its 2315-mm (91.1-in) wheelbase is 15 mm (0.6 in) shorter. Front and rear overhangs have each been reduced by 90 mm (3.5 in).

Overall, the ND Miata is 50 mm (2.0 in) shorter than the original NA. The smaller dimensions along with a thorough attention to mass reduction helped bring the base curb weight down by 150 lb (kg) to 2332 lb (1058 kg), which helped improve fuel efficiency by 25% compared with the outgoing model.

Miata has never been a vehicle for drivers outside of the 90th percentile, but the 2016 cabin's admittedly snug (for this 6-ft, 3-in *Automotive Engineering* editor) space feels slightly roomier than that of the previous car. While this still isn't a vehicle for drivers with size-12 feet, there's a bit more generous driver's seat recline, the steering wheel tilts (but doesn't telescope) and the seats are lower by almost 1 in (25 mm), which provides about 10 mm (0.40 in) more headroom.

Equipping the 2.0-L engine with a low-profile oil pan and positioning the powertrain about 12 mm (0.5 in) farther back in the chassis enabled the hood's outer surface to be more than an inch lower than the NC car's, according to Dave Coleman, the

**Fourth-gen MX-5 is underpinned by a significantly optimized version of the original NA car's ladder-type backbone chassis. In the 2016 version, the rear suspension crossmember is located underneath the differential rather than above it as on the previous NC Miata. That puts the crossmember closer to the chassis-load input vector and creates a simpler, lighter arrangement. Upper-and-lower control arm suspension continues with revised geometry and aluminum wishbones and new steering knuckles. Rack-mounted electric-assist power steering makes its first appearance on the 2016 MX-5. Rear 11-in brake rotors are solid, not vented.**

program's Lead North American Development Engineer. Forward visibility is outstanding as a result.

During a drive of early production cars in Michigan last spring, Coleman explained that retaining the industry's easiest-to-operate (one hand!) manual fabric convertible top was a program mandate. "Although the 2016 top is fully insulated, it's 65 grams lighter than the previous top thanks to new aluminum top bows and structure," he said. Our drives in the car indicated that the new top is also quieter.

## Less is more in power

Yamamoto's engineers had to be aggressive in their hunt for excess mass. "Every gram mattered," noted Coleman, "because the NC car was not exactly overweight by most standards." The ND's high-tensile—780 MPa (113 ksi) or higher—steel content increased to 71% versus 58% on NC. All exterior panels except the windshield frame and doors are made of aluminum alloy, which helped reduce the body-in-white mass by about 45 lb (20 kg), Coleman said.

The new seats are each 17 lb (8 kg) lighter than before, and 35% thinner due to the previous seat springs replaced by a net and urethane cushion. The new A/C system is approximately 20% lighter—8 lb (3.6 kg)—than that on the NC Miata. The long list of redesigned components and their mass savings includes the suspension system—26 lb (11.8 kg)—much of it from replacing the previous steel steering knuckles with aluminum ones; manual transmission—16 lb (7.3 kg); front 11-in (280-mm) brake rotors—14 lb (6.3 kg); and instruments and controls—3 lb (1.4 kg). Fastener count was reduced significantly through component design and consolidation, Coleman said.

Reducing curb weight enabled Mazda engineers to rethink the MX-5's powertrain specs. On the new car, 155 hp (115 kW) at 6000 rpm does the job of 167 hp (124 kW) at 7000 rpm on the previous model, both using 2.0-L naturally aspirated inline four-cylinder engines. Coleman explained that the long-stroke—83.5 x 91.2 mm (3.29 x 3.59 in)—Skyactiv-G unit delivers more mid-range power, while torque—148 lb·ft (200 N·m) at 4600 rpm, versus 140 lb·ft (190 N·m) at 5000 rpm on the NC's engine—is also stronger and delivered at a more useful engine speed.

## Three feature bundles

The U.S. EPA fuel economy rating was not yet finalized when this article went to press, but should be close to 27 mpg city/34 mpg highway/30 mpg combined when equipped with the standard six-speed, Mazda-built manual transmission (6th gear is direct, not overdrive). The Aisin six-speed automatic models offer an EPA-estimated 27 mpg city/36 mpg highway/30 mpg combined. Only the 2.0-L engine is available in North American market Miatas; a 1.5-L unit serves as the base engine in most global markets.

**MX-5 cockpit, though tight for big-and-tall occupants, is among the industry's most simple and distraction-free driving environments. But the vertical navi screen is an awkward solution and looks like an afterthought.**

And in furthering MX-5's worldwide popularity as a racecar, Mazda is offering the 2016 MX-5 Cup version as unveiled at the SEMA show (see http://articles.sae.org/13697/). Street models are now available in three trim levels with limited options. The base Sport model comes standard with 16-in wheels, cloth top, and LED headlamps (with halogen DRLs). Next step up is the Club, with 17-in wheels, LED DRLs, shock tower brace, limited-slip differential, Bilstein dampers with tighter valving, special air dams and rear spoiler, and Bose audio. The Club trim also offers BBS alloy wheels and Brembo foundation brakes.

Top of range is the MX-5 Grand Touring. It includes leather seat surfaces, insulated headliner, and a more sophisticated Bose audio system.

Overall, Mazda's ND development team has done an outstanding job in keeping the new MX-5 true to its roots while increasing the car's efficiency; improving its handling, steering and passenger comfort; and crafting a generally handier package for everyday use. And luckily for FCA, there's more than a little bit of MX-5 in every 124 Spider! ■

*Lindsay Brooke*

# Mazda trickles out more new MX-5 details

**New MX-5 is lighter, lower, shorter, and wider than before.**

The new Mazda MX-5 made its first public European appearance at the 2014 Paris Motor Show, bringing a lighter car than the outgoing model, powered by a new 1.5-L gasoline engine (the U.S. and possibly other markets will get a 2.0-L engine, according to industry reports). Mazda is giving few powertrain details away at this stage, but a six-speed manual transmission will be offered. Mazda's SkyActiv technologies are used in both vehicle design and construction as well as the engine.

Compared with the outgoing model, the new MX-5 is 105 mm (4.1 in) shorter, 20 mm (0.8 in) lower, and 10 mm (0.4 in) wider. The wheelbase is also shorter by 15 mm (0.6 in) at 2315 mm (91.1 in), while both front and rear overhangs have been reduced by 90 mm (3.5 in). Overall, the car is 50 mm (2.0 in) shorter than the first generation MX-5.

Not surprisingly, Mazda has retained the existing format for the car with front mounted engine and transmission and rear-wheel-drive. Similarly the two-seat open bodywork is retained.

With driving the sharp focus for the MX-5, the Paris Show booth included a chassis assembly of front and rear suspension with engine and transmission and bracing section between transmission and rear axle. There are many similarities to the technology used on the outgoing model, but the components are new.

"The format of the rear suspension is the same as the third generation, the multilink, but what has

**New MX-5 is 105 mm shorter than its predecessor and 50 mm shorter than the original. Wheelbase is shorter too.**

changed is the position of each of the links", Mazda MX-5 Program Manager Nobuhiro Yamamoto told *Automotive Engineering*. "I can't discuss the geometry, but the suspension crossmember goes underneath the differential, while on the third-generation model, it was fitted over the top. The reason is quite simple. It's going back to the fundamentals, because the crossmember has to be reacting to the input that's coming from the links and it needs to provide that rigidity. So rather than going over the top, if the crossmember goes underneath, which is closer to the vector where the forces are coming in, it makes it simpler, you can make it lighter, yet it will still give enough rigidity. SkyActiv technology is going back to basic principals, and that's what gives us the ideal form."

Mazda has continued with double wishbone suspension at the front: "At first glance, it might not look as though there have been changes, but there are," said Yamamoto. "As with the rear, we have changed the geometry, and it's important to make the unsprung weight as light as possible, so the upper and lower arms are made from aluminum.

"The steering knuckle on the third-generation car was made from steel, but on the new car, it's made from aluminum, so it's lighter but still has high rigidity. Up to the third generation, we used what we called a "pancake" bush for the lower arm rear, mounted through the bush from underneath. It was basically made for a front-wheel-drive vehicle, but we looked at it and said we need to go back to first principals. If you look at the compliance, the movement of the lower arm, we said that we need to make it into a roll-cage shape. It's a matter of looking at the ideal and working out the simplest way to get at it, and this is the basis of all these different changes."

Power assistance for the steering is now a completely electrical system, whereas it was previously hydraulic: "You can see that the power assistance motor is mounted on the rack," continued Yamamoto. "But if you have a look at all the other vehicles in the Mazda lineup, and probably others' as well, it's more common to fit the motor on the steering column. You have to have precise feedback in a sports car, and for us to achieve that the motor has to be on the rack."

The cabin has been moved slightly further back, and the hip points of the seats have been lowered, reflecting the lower center of gravity of the new car. ■

*John Kendall*

**MX-5 chassis now features a rear axle with underslung crossmember. Aluminum steering knuckle replaces steel.**

# Mazda takes wraps off 2016 MX-5Miata

**The all-new MX-5, which adopts both SkyActiv technology and Mazda's Kodo design language, will be rolled out to global markets starting in 2015.**

Mazda Motor Corp. on Wednesday evening unveiled the fourth-generation MX-5 Miata, which adopts both SkyActiv technology and Mazda's "Kodo" (or "Soul of Motion") design language. The MX-5, first introduced 25 years ago, maintains Mazda's original aim of offering the pure driving fun that only a lightweight sports car can provide. Weight reduction of more than 100 kg (220 lb), down from the previous model's 2480 lb (1123 kg), is achieved through the adoption of SkyActiv technology—a combination of a lightweight and rigid chassis and body and of highly efficient engines and transmissions. This MX-5 chassis is the first time SkyActiv Technology has ever been shown in a front mid-ship engine rear-wheel-drive design. Its compact configuration showcases an engine closer to the vehicle's center, a center of gravity that itself is lower than that of any previous generation. "The dedication of the R&D and Design teams to honor the vehicle's dynamic heritage and globally recognizable features is evident from the chassis' structural blueprint and will undoubtedly translate into what we believe will be a remarkable production car," said Robert Davis, Senior Vice President, U.S. Operations, Mazda North American Operations. According to Mazda, this is the most compact MX-5 thus far, while also achieving the highest standards of body rigidity and collision safety performance. The vehicle will begin being rolled out to global markets starting in 2015. ■

*Matthew Monaghan*

SAE Eye on Engineering Quick Take:

## Mazda's new Miata

Mazda is updating its sports car icon. In this episode of SAE Eye on Engineering, Senior Editor Lindsay Brooke looks at the new MX-5 Miata. SAE Eye on Engineering can be viewed at https://youtu.be/i8h7J0aEOdY. It also airs in audio-only form Monday mornings on WJR 760 AM Detroit's Paul W. Smith Show.

Access archived episodes of SAE Eye on Engineering at www.sae.org/magazines/podcasts.

# McLaren debuts new entry-level Sports Series

**The McLaren 570S boasts an impressively low 2895-lb curb weight thanks to its carbon fiber chassis and aluminum bodywork.**

McLaren Automotive debuted a new entry-level model for its family of super sports cars at the 2015 New York Auto International Show, with the 570S introducing a new carbon-fiber monocoque chassis for the company that has larger door openings for more convenient ingress and egress.

The 570S is the first member of a product line McLaren brands its Sports Series, which represents an entry level alongside the company's existing Super Series and Ultimate Series models. These represent a hierarchy in the company's line, demonstrating seriousness of sporting intent and price.

Of course, practical larger openings compromise rigidity, so the chassis is reinforced with some additional material to preserve its integrity. Even with that strengthening, the bare monocoque weighs only 80 kg (176 lb), while it is a claimed 25% stiffer than competitors' aluminum chassis.

The 570S's doorsills are 80 mm (3.1 in) lower than on the company's earlier models, and are also thinner, for a less cumbersome entry in and out. The car can be equipped to weigh as little as 1313 kg (2895 lb), which McLaren says is 150 kg (331 lb) less than its lightest competitor. The 570S also has thinner B-pillars for improved visibility.

"For the Sports Series we wanted to offer a real sports car feel," explained Chief Test Driver Chris Goodwin. "Ultimately, it means a car then feels nimble, agile, and it must also possess a lightness

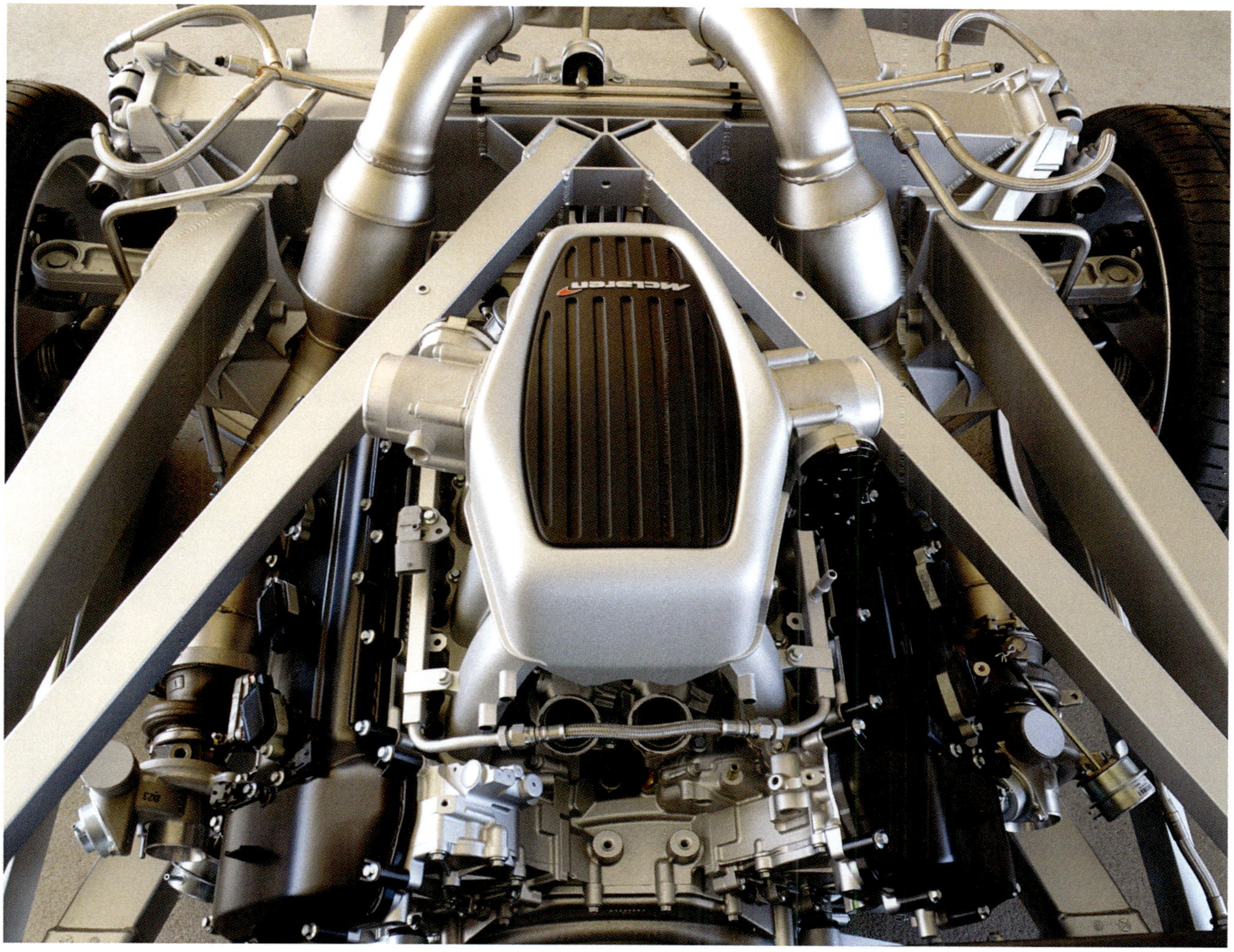

**The 570S's new M838TE twin-turbo 3.8-L V8 is revised from this earlier engine in the 650S to include upgraded cam phasers.**

and directness of response to drivers' inputs. Certainly for us, a sports car really needs to engage the driver at any speed."

Naturally, tires contribute significantly to that mission. For the 570S, McLaren's goal was to provide useable performance with approachable limits, rather than maximum performance with a precipitous drop-off beyond the limits. To that end, Pirelli partnered with McLaren to tune the 570S's P Zero Corsas with driver-friendly characteristics.

"We have worked closely with Pirelli throughout the program to design a bespoke tire to best suit the Sports Series, and it is a really important partnership," Goodwin noted. "Being the only contact with the road, the tire is critical to how the car handles and behaves, and we treat them in the same way as any key suspension component," he added. "They are specially tuned and developed with Pirelli, and that makes a huge difference to the dynamics of this car."

Tire sizes are 225/35 R19 on the front and 285/35 R20 on the rear. Regular P Zero tires are a no-cost option for drivers seeking longer tire life and softer ride, while the standard P Zero Corsa's stiffer sidewalls and grippier tread boosts acceleration, braking, and lateral acceleration for improved track times. The tires mount on 19-inch front wheels and 20-inch rears.

Beneath the engine cover, McLaren's new M838TE 3.8-L twin-turbo V8 engine is similarly revised, with 30% new components compared to the M838T. It is now rated at 570 PS (562 hp/419 kW) as referenced in the car's name and 600 N·m (443 lb·ft). It propels the car to 100 km/h (62 mph) from rest in just 3.2 s and to 200 km/h (124 mph) in 9.5 s. Terminal velocity is 204 mph (328 km/h).

The changes to the engine include installation of an automatic engine stop/start system for improved fuel efficiency that produces 25.2 mpg on the combined European driving cycle and 258 g/km of CO2. The car will escape U.S. gas guzzler taxes.

New 60-degree cam-phase controllers reduce internal engine inertia and improve control, which helps both efficiency and throttle response. New equal-length stainless steel exhaust manifolds are mated to hydroformed steel tubes to optimize exhaust flow while enhancing the car's engine note.

**An added console switch gives the driver direct control of the improved traction control and electronic stability control systems.**

As in previous models, the engine continues to be fitted with a dry-sump oil system and a flat-plane crankshaft. The 570S carries over the same seven-speed dual-clutch transmission seen on the 650S.

McLaren's programmers have been busy improving the company's software, so the 570S enjoys new anti-lock brake software that provides better brake pedal feel and reduces the pulsing the driver feels through the pedal when the ABS is active.

New electronic stability control, developed in conjunction with hardware supplier Bosch, adds a Dynamic mode for reduced computer interference with the driver to permit increased ability to slide the rear of the car in a drift while the ESC is active.

For the first time, the ESC system has its own control switch, so drivers can independently set its mode; full, dynamic, or off. Bosch upgraded the pump and dual-valve system for increased resolution so that the 570S enjoys more precise, progressive ESC engagement for less intrusion.

The traction control is also reprogrammed, with more precise control in extreme situations such as drifting. It may be rare for owners to exploit these new capabilities, but they should contribute to some spectacular Internet videos. ■

*Dan Carney*

# Mercedes-Benz replaces M-Class with GLE, adds plug-in hybrid

**Mercedes-Benz GLE series is a partly refreshed lineup with restyled front and rear.**

The new Mercedes-Benz GLE SUV, introduced at the 2015 New York Auto Show and scheduled for market in late summer, could be regarded as just a major "refresh" of the M-Class. Four of the powertrains are carryovers, and the body styling is primarily different in the front and rear styling treatments. So externally it looks very different only from the recently shown GLE five-door "coupe," which has a specific body, but with which it shares platform, mechanical components, and most of the interior. Yes, all the familiar crossover models are back, although with new designations based on the GLE name (GL for SUV and the third letter for the platform).

Considering the lineup mostly a set of renamed returnees would be an unfair assessment, as the 2016 edition is on a stretched wheelbase—from 114.8 to 121 in (2916 to 3073 mm), although overall length is barely an inch longer at 190 in (4826 mm).

## Plug-in hybrid

There's a significant new model in the range: the 550e 4Matic, a plug-in hybrid (PHEV). It is a first for the line and will be available later in the year. It features a version of the carryover 3.0-L V6, the twin-turbo rated at 329 hp (245 kW) and 354 lb·ft (480 N·m). The electric motor, the same as used in the S-Class hybrid, is listed at 85 kW and 251 lb·ft (340 N·m).

Peak torque from engine and motor is 479 lb·ft (649 N·m), which produces what was described as "impressive" acceleration, comparable to a V8 with the fuel economy of a four-cylinder. The GLE 550e has the ability to accelerate up to 81 mph (130 km/h) on the electric system alone. The lithium-ion battery pack has a capacity of 8.8 kW·h and provides an all-electric range of about 19 mi (30 km).

Although U.S. fuel economy figures aren't yet available, Mercedes-Benz is claiming the PHEV will deliver better fuel economy than the 2.1-L diesel with its seven-speed automatic (a nine-speed automatic is available in Europe). The seven-speed 7G-Tronic is used in all U.S. market GLE models except the AMG series, although the one in the PHEV is a much-modified version.

A center console switch permits the driver to select from four modes, which are commonly available on PHEVs: Conventional Hybrid, E-mode for all-electric operation, E-save to maintain the charge for later use, as in an area where gasoline-engine operation might not be permitted; and Charge, which charges the battery pack when driving and at engine idle.

An intelligent drive management system promotes anticipatory driving, particularly deceleration, to enhance regenerative braking. It integrates such information as knowledge of the road conditions ahead (using GPS and radar sensors) to adjust the powertrain operating parameters for optimum energy consumption.

The twin-turbo V6 air intake and the motor electronics share a low-pressure liquid cooling system separate from the engine cooling.

The GLE-550e is the latest PHEV in the Mercedes-Benz line. The S-Class was the first and a C-Class PHEV is coming in the fall. The company has said it expects to have a total of 10 plug-ins by 2017.

**The GLE plug-in hybrid has low-temperature liquid cooling system. Note reservoir just visible under engine cover. System cools air intakes of the twin turbos and the motor electronics.**

## Carryover engines

The base 350 GLE has the 3.5-L direct-injection V6 rated at 302 hp (225 kW). It is the only model available in a rear-drive-only configuration; all others have 4Matic permanent four-wheel drive. The twin-turbo 3.0-L V6 model, formerly called the ML400, is now the GLE 400. As in the PHEV, it is rated at 329 hp (245 kW). For the 2016 models, this direct-injection engine incorporates piezo-electric injectors.

The diesel-powered model, which had been called the ML250 BlueTEC, is now the GLE 300d 4Matic. The engine is the 2.1-L four-cylinder rated at 201 hp (150 kW) and 369 lb·ft (500 N·m). With the nine-speed transmission in Europe, the diesel is the first SUV in its size class with a "conventional drive" (non-hybrid or EV) to produce emissions of just 140 g CO2/km on the NEDC (New European Drive Cycle). Of course, the PHEV does even better, producing just 78 g CO2/km.

The 4Matic system includes Dynamic Select, the road terrain program with a console selector knob. It offers comfort, slippery surface, individual selection, and sports modes, with a program that modifies anti-lock brakes response and traction/stability controls.

Optional Airmatic air suspension is back, retuned for the new models and combined with ADS (automatic damping) using a strategy based on the Dynamic Select control position. Airmatic/ADS can be combined with Mercedes-Benz Active Curve System, which uses motor-operated anti-sway bars in front and rear to reduce body roll. Active Curve is standard on the AMG models.

**With the GLE's hood raised, it's apparent the hood grilles feed air into the cabin air intake on each side, although on the driver's side the flow is through a port.**

## AMG at top of lineup

The AMG series is back and at the top of the lineup with two models, the GLE63 4Matic and the GLE63 S 4Matic. They both have the 5.5-L V8 newly rated at 550 hp (410 kW) in the base model and 577 hp (430 kW) in the S. Those ratings reflect significant retuning from the 518 hp (386 kW) of the previous model.

The AMG series is offered with three other exclusive driving features. One is the nine-speed automatic, which is standard. Second is an optional Off-Road Engineering Package, which adds an Offroad+ position to the Dynamic Select. It contains an off-road reduction gear and 100% inter-axle differential lock. Third is the combination of the Off-Road Engineering Package with the optional AirMatic air suspension that provides up to 11 in (279 mm) ground clearance and a fording depth of 23 in (584 mm). It incorporates a reinforced underfloor panel to protect underbody components on extreme driving surfaces.

Appearances notwithstanding, the air intakes on the hood of all models are not for engine breathing. Rather, they provide outside air for the passenger compartment.

The GLE carries over the full suite of Mercedes-Benz safety-collision and auto-braking systems, and adds Crosswind Assist and Collision Prevention Assist-Plus. The Plus system warns the driver if he/she is not maintaining an appropriate distance from the vehicle ahead and if the driver doesn't respond will brake the vehicle to prevent the possibility of collision.

The vehicles will continue to be produced at the Mercedes-Benz plant in Alabama. ■

*Paul Weissler*

# Mercedes-Benz melds SUV and coupe styling for GLE Coupe

**Mercedes-Benz' new GLE Coupe range spans a V6 diesel to a 270-kW 450 AMG version.**

When is an SUV not an SUV? When it's the new Mercedes-Benz GLE Coupe, a question and answer that could equally apply to BMW with its M6, and an indication of a global trend towards sporty SUVs with highly sophisticated engineering and systems, wrapped in a bodyshell that might appeal to buyers less than 50 years of age.

Based on the GLE (which was the SUV in the E-Class segment formerly designated M-Class), the GLE Coupe is due to be publicly unveiled at the 2015 North American International Auto Show in Detroit. The top variant, the gasoline 270-kW (362-hp) GLE 450 AMG 4-Matic, is described by Mercedes as introducing a new Mercedes-AMG product line: "the AMG sports models."

There are those who might consider that already more or less exists, but apparently not in quite the way Mercedes wants. It's subtle stuff and strengthens the AMG message.

However, there is little subtle or understated about the (to European eyes), bruiser of a cruiser that is the GLE. The 450 AMG sits on a 2915-mm (114.8-in) wheelbase, with a length of 4900 mm (192.9 in), width of 2003 mm (78.9 in), and height of 1731 mm (68.1 in). Its 3.0-L V6 engine delivers 520 N·m (384 lb·ft) of torque from 1400 to 4000 rpm, drives through a 9G-Tronic (nine-speed) automatic transmission, and puts all this on to the road (or track, or field, or muddy morass), via optional 22-in wheels and a 4-Matic permanent all-wheel drive (AWD) system. Front/rear bias of the 4-Matic AMG version is 40:60; lower powered GLEs get a 50:50 split.

A 245-kW (329-hp) version of the bi-turbo engine is available in the GLE 400, and there is also a 3.0-L V6 diesel producing maximum torque of 620 N·m (457 lb·ft) from 1600 to 2400 rpm.

Successfully melding two model classes, an SUV with a coupe, which is what Mercedes says it has done with its new arrival is not easy, and the company's descriptive phrases such as "the engine

**The distinctive rear-end treatment of the GLE Coupe. An opening panoramic glass sunroof is an option.**

produces the whole range of sounds from sporty and dominant to poised and reticent", and the car "can create a superior feeling of calm, coupled with unshakably serene driving dynamics", sound a little quaint and out of kilter with the company's advanced design and engineering message.

Although based on the GLE (M-Class) architecture, much of that has been subjected to extensive development to bring mechanical reality to the sobriquet "coupe."

Significant in this are the chassis and transmission electronics. Via a Dynamic Select system, the driver can choose, using via a rotary control, five settings: Individual, Comfort, Slippery, Sport, and Sport+.

"The system networks all the control systems, generating different sets of parameters: Sport and Sport+ turn the GLE 450 AMG into a sporty, dynamic vehicle with pronounced longitudinal and lateral dynamics," states Mercedes. Comfort gives sedan-like cruising with energy efficiency, and the sensibly termed "Slippery" speaks for itself. As with some other Mercedes' models, the Individual setting facilitates driver programing for personal control strategies.

Dynamic Select looks after powertrain, suspension, and steering settings and also what Mercedes terms, again rather curiously, "soft skills." These include engine sounds, instrument displays, and "dynamic control of the driving lights"—the active light function of the all-LED headlights.

In Sport+, gear shifting is described as being "acoustically accentuated," with downshifting making a double-declutch sound, and the car's large 8-in central Comand TFT display provides graphics of longitudinal and lateral acceleration plus steering-wheel angle via a photo-realistic motorsports graphic.

**Melding SUV and coupe profiles: the Mercedes-Benz GLE Coupe.**

**Interior of the Mercedes-Benz GLE Coupe is said to be driver-centric.**

Depending on model and specification, the suspension offers steel springs or Airmatic systems.

Airmatic is available for the first time with a continuously variable damping system called ADS Plus, operating in combination with Dynamic Select.

Operating conditions are monitored via a steering-angle sensor, four turn-angle sensors, the driving speed from the ESP signal, and information from the brake pedal switch. The control unit uses the signals to infinitely adjust the damping force at each wheel. Sport Direct-Steer is fitted with a ratio of 17.4:1.

The car also has Mercedes' Active Curve System, which uses active roll control when Sport or Sport+ is selected. Active anti-roll bars are fitted front and rear, and are said by Mercedes to compensate successfully for asymmetric effect such as striking a pothole.

The lower powered GLE 400 4-Matic and diesel GLE 350 d 4-Matic come standard with Agility Control. Coil spring steel suspension is standard, with an amplitude selective damping system. Airmatic is an option with ADS Plus.

The GLE Coupe's styling carries some of the visual cues of other Mercedes models, including the latest generation C-Class, which will also appear as an SUV in 2015. The rear of the car is influenced by design lines of the S-Class coupe with slim lights.

The car's glasshouse is relatively narrow, with a belt line gently rising towards the rear. The rear window is rounded off at the top, which is a long established Mercedes' signature on load-carrying models.

The nose of the car carries the three-pointed Mercedes' star with single central chrome louver.

Interior of the car can be lightened by an optional opening glass panoramic glass roof. AMG sports seats are fitted in the top version. At 1650 L (58.3 ft$^3$), the GLE offers the largest luggage compartment in its class, claims Mercedes. A fully electric extendable and retractable trailer coupling, which includes ESP trailer stabilization, is offered. The GLE can tow a 3500-kg (7700-lb) braked trailer.

With the launch of the GLE, Mercedes has clarified its model range nomenclature. The five core model series are A, B, C, E, and S. All SUVs will now have GL in their name. So the GLA is the SUV in the A-Class segment; GLC is the SUV in the C-Class segment that was formerly GLK; GLE is the SUV in the E-Class segment that was formerly M-Class; GLE Coupe is the SUV in the E-Class Coupe segment; GLS is the SUV in the S-Class segment that was formerly GL. The long-established G Wagen remains unchanged. If you are still confused, talk to your local dealer.

Tobias Moers, Chairman of the Board of Management of Mercedes-AMG, is definitely not confused: "The new product line from Mercedes-AMG has allowed us to make true sports car technology and the fascination of motorsports more accessible." ■

*Stuart Birch*

# Most powerful production Mini shows in Detroit

**Power up by 10% and improved aerodynamics give the new generation Mini John Cooper Works a 245-km/h top speed.**

When the first BMC Mini was introduced in 1959, its power output was 25 kW (34 hp) from a very modest 848 $cm^3$ engine that could eventually take it to 100 km/h (62 mph) in about 26 s and on to a top speed of 120 km/h (75 mph)—on a good day.

At the 2015 North American Auto Show in Detroit, the most powerful production Mini ever will be revealed. Its power output is 170 kW (228 hp) from a still relatively modest 2.0-L turbocharged engine that can take it to 100 km/h in 6.1 s and on to a top speed of 245 km/h (152 mph), more than doubling that of the original car.

The new Mini is the latest generation of the John Cooper Works (JCW) range leader. Compared to the outgoing 1.6-L model, the new 2.0-L presents a power hike up by 10%, and maximum torque, at 320 N·m (236 lb·ft) from 1250 to 4800 rpm, improved by a very impressive 23% to enhance driving flexibility. Acceleration from 80 to 120 km/h (50 to 75 mph) is down by 10%.

The car's engine, based on the Mini Cooper S unit, uses fully variable valve control and variable camshaft control. Improved aerodynamic efficiency has also contributed to the performance gains, and Mini has also focused on minimizing weight without losing the car's premium equipment.

Transmission options are a six-speed Steptronic or six-speed manual, the former giving the 6.1-s 0-100 km/h time, the latter increasing this slightly by 0.2 s.

Mini describes the latest JCW as drawing on the brand's motorsport expertise with a package of suspension, brake, aerodynamic, and interior modifications "to deliver race-car sensations without losing sight of Mini's premium position."

**Aerodynamic improvements for the John Cooper Works include a model-specific rear apron and rear spoiler. Chromed tailpipes are centrally mounted and help provide the required aural signature.**

Or its emphasis on fuel consumption. The original Mini averaged between about 7.0 to 5.7 L/km. The JCW's combined NEDC consumption with the manual gearbox is 6.7 L/100 km with CO2 emissions of 155 g/km. With Steptronic, economy improves by 20% to 5.7 L/100 km (coincidentally almost exactly the same as the original Mini's best), with CO2 emissions down to 133 g/km. The original Mini weighed 585 kg (1290 lb), around half that of the JCW.

Both the suspension and the ZF Servotronic electromechanical power steering have been tuned to complement the JCW's extra power and torque. Newly developed four-piston Brembo brakes within 17-in light-alloy wheels are fitted; 18-in are optional. Dynamic Stability Control (DSC), Dynamic Traction Control (DTC), Electronic Differential Lock Control (EDLC), and Performance Control (PC) are standard, with Dynamic Damper Control an option.

Mini stresses the JCW's aerodynamic changes have been introduced "with a purpose rather than for purely aesthetic reasons." They include a new front apron with large engine air intakes, modified side sills, a model-specific rear apron, and a JCW spoiler.

The car has LED headlights with white turn indicators, wheel-arch surrounds, and unique radiator grille, side scuttles, and tailgate with JCW badging.

Compared to other Mini versions, the JCW gets exclusive to model bucket seats with integrated headrests in Dinamica/fabric and a multi-function steering wheel with shift paddles (in conjunction with Steptronic transmission), gear selector, instrument displays, stainless steel pedals, and footrest.

Bespoke JCW personalization options include a model-specific head-up display (HUD), Chili Red roof and door mirror covers, bonnet stripes, and exterior and interior John Cooper Works Pro design accessories in addition to the other options available for the entire Mini range. ■

*Stuart Birch*

# Mitsubishi puts NVH in 2016 Outlander's crosshairs

**All-new sheet metal from the windshield forward and more than 100 enhancements define the 2016 Mitsubishi Outlander, a crossover utility vehicle that has received the Insurance Institute for Highway Safety's Top Safety Pick+ each year since the 2014MY.**

Two years after the third-generation Mitsubishi Outlander arrived in the U.S. market, the crossover utility vehicle was revised for the 2016 MY with new signature front-end styling and 100-plus engineering and design changes.

"It was very clear to us that (certain) things, like exterior styling and drivability, [were] below the competitive standard, [and] had to be changed very quickly," Ken Konieczka, Vice President of Sales at Mitsubishi Motors North America (MMNA), said at a recent San Francisco, CA, media ride and drive program for the 2016 Outlander.

The latest Outlander is the first MMNA vehicle to showcase the front fascia Dynamic Shield styling, and many of the seven-passenger, three-seating-row vehicle's technical modifications are aimed at refining the driving experience. For instance, the next-generation CVT (continuously variable transmission) reduces the transmission torque loss by 26%, effectively improving the amount of power transferred to the wheels.

**The 2016 Mitsubishi Outlander is significantly quieter than its predecessor, according to Mitsubishi's Melvyn Bautista, who is shown in front of the 2016 model.**

This new CVT-8 and a recalibration of the transmission ECU elicit improvements in both the standing start and passing acceleration. In all-wheel-drive models equipped with the 2.4-L four-cylinder engine and CVT, 0 to 60 mph (97 km/h) acceleration improved by 1 s, going from 11.2 to 10.2 s.

Mitsubishi's CVT-8 debuted in the U.S. on the 2015 Outlander Sport equipped with the 2.0-L four-cylinder engine. On the 2016 Outlander, the new CVT has a gear spread from 2.631: to 0.378:1 and a final drive ratio of 6.026:1. Outlander's previous CVT had a gear spread from 2.349: to 0.394:1 and a final drive ratio of 6.466:1.

Ride and handling modifications were made to the MacPherson strut front and multi-link rear suspension. Those changes include front suspension crossmember reinforcement, front shock absorber re-tuning, rear suspension spring rate and damping changes, as well as larger shock absorbers in the rear. The re-programming of the electric power steering also helps improve the 2016 vehicle's ride and handling.

Outlander's suspension, engine and transmission cradles, body panels, side mirrors, windshield, and rear glass were the source points for 31 noise-reduction solutions.

According to Melvyn Bautista, MMNA's Product Strategy Manager, the various sound-absorbing techniques were undertaken based on comments from U.S. customers, JD Power Initial Quality Survey results, and third-party static and dynamic vehicle evaluations.

The noise reduction crusade even had engineers from Japan placing microphones throughout the current generation Outlander's suspension. "Six hours a day for weeks they drove the vehicle back and forth over a rough residential street near MMNA's Cypress (CA) headquarters to get data and come up with solutions to minimize the wind and road noise going into the cabin," Bautista said in an interview with *Automotive Engineering*.

Noise reduction is notable in the low-frequency range. For instance, compared to the 2015 Outlander, the 2016 vehicle's front center noise was reduced from 64.1 to 60.8 dB in the 10-500 Hz frequency range.

The 105.1-in (2670-mm) wheelbase Outlander's new offerings include LED headlamps, LED running lamps, LED taillamps, power folding side mirrors, and a windshield wiper de-icer. "It's an all-new feature for MMNA," said Bautista, "An electric heater element, located underneath where the wipers sit, helps prevent the windshield wipers from sticking in the park position."

Outlander is available with the 166-hp (124-kW) four-cylinder engine mated to the CVT or a 224-hp (167-kW) 3.0-L V6 engine mated to a six-speed automatic transmission. The four-cylinder-powered front-wheel-drive model has EPA mpg ratings of 25 city/31 highway, while the AWD version's mpg is 24 city/29 highway. The V6-powered AWD model has mpg ratings of 20 city/27 highway.

The 2016 Outlander goes on sale in July 2015 with a MSRP of $22,995 for the base FWD model. ■

*Kami Buchholz*

# Mitsubishi refreshes Outlander

**Restyled front (and rear) of the 2016 Outlander reflects new Mitsubishi design language.**

Mitsubishi is working to bring the brand into the mainstream and on the shopping lists of more buyers, and the 2016 Outlander seven-passenger, three-seat row crossover, introduced at the 2015 New York International Auto Show, represents a major undertaking. It will go on sale at the end of the year.

Compared to its predecessor, body dimensions see no big differences; the 105.1-in (2670-mm) wheelbase is the same, and the 184.8-in (4694-mm) overall length is a mere 1.5-in (38-mm) increase from 2015. However, the Outlander exterior was a "clean sheet" vehicle just three years ago, and what it needed was a front-to-rear effort to create a more appealing driving experience. Virtually all the weight reductions were taken in the previous model, and the 2016's more than 100 changes were made with moderate total increases 11-22 lb (5-10 kg) in vehicle weight.

## Emphasis on quiet cabin

The front and rear of the new Outlander were restyled and now reflect a new design language. Delivering a quality experience is where Mitsubishi invested. Under the skin are structural upgrades to improve body rigidity, the suspension was retuned front and rear, and electric power steering was reprogrammed for ride/handling.

The vehicle already had a Top Safety Pick+ rating with the previous model. It had earned a good rating in each of the Insurance Institute for Highway Safety's five categories, including the small overlap front crash test that has been the most challenging.

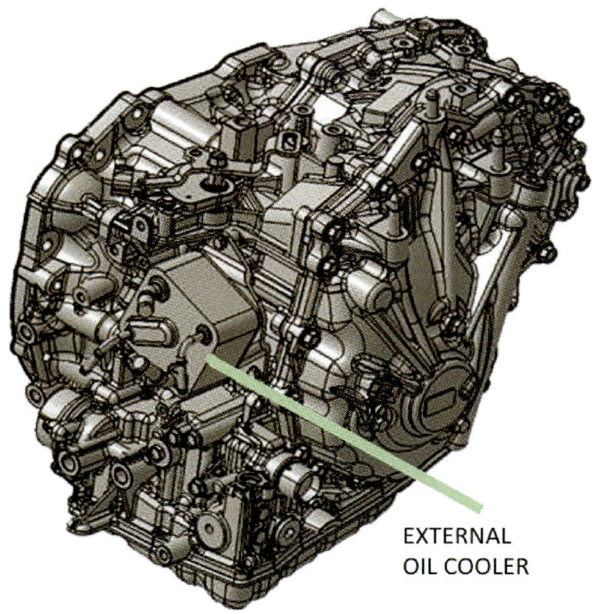


**CVT was reworked, increasing overall ratio spread and reducing friction. Oil cooler was moved from interior to external mount.**

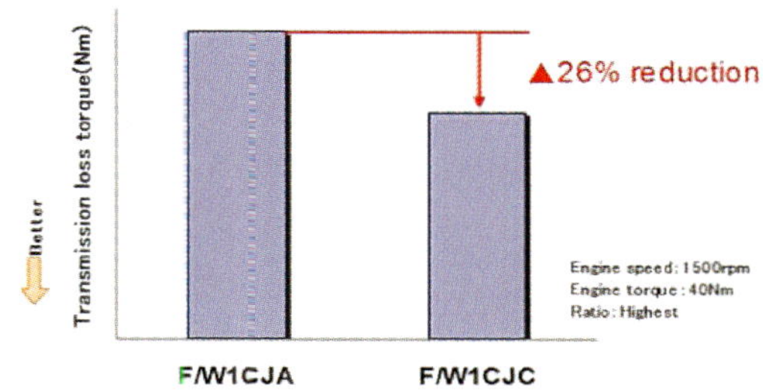


**Graph shows 26% reduction in torque loss from anti-friction improvements in the 2016 CVT (F/W1CJC) from 2015 CVT (F/W1CJA) under test conditions listed (1500 rpm, 40 N·m torque throughput and highest pulley ratio).**

Forward-collision mitigation and lane-departure warning are in an option package for the 2016 SEL and GT trims.

Making the Outlander quieter was a priority and the vehicle has a range of additions, from more sound insulation throughout, to improved door weatherstripping and body sealing, and a thicker, insulating windshield and rear door glass.

Major technical changes were made to the standard CVT (continuously variable transmission), which is used on all trim levels with the 2.4-L four-cylinder. The internals of the CVT were reworked, so the oil-to-water cooler was relocated to the exterior.

The CVT's belt and pulleys were redesigned with a primary objective to increase the ratio spread. A good ratio spread is around 6.0, and many CVTs are under that number, but the 2016 Mitsubishi CVT is at almost 7.0. The greater the spread, the more flexibility in calibrations for acceleration and fuel economy. The new spread is from 2.631: to 0.378:1, with a final drive ratio of 6.026:1. This compares with 2015 numbers of 2.349: to 0.394:1 and a final drive ratio of 6.466:1. With these new ratios, the launch ratio can be numerically higher for better acceleration and the cruise/low-load ratio can be numerically lower for better fuel economy.

The four-cylinder engine produces 162 lb·ft (220 N·m), which the CVT can handle. The newest design belt-drive CVT for a performance V6 of a competitive Japanese make has a ratio spread of 6.3, and that represented an increase from 5.4. However, Mitsubishi's 3.0-L V6 for the Outlander, which is combined with all-wheel drive uses a conventional six-speed automatic.

## CVT friction reduction

Another engineering objective was to reduce CVT friction, also for better fuel economy, so one aspect of the steel link belt redesign was for reduced friction. Furthermore, the oil pump was downsized, line pressure was reduced, and the internal baffle was redesigned to minimize oil churning, additional steps that improve CVT efficiency. The overall result was a 26% drop in transmission torque loss with the new design.

The new CVT weighs just 91.8 lb (41.6 kg) wet, compared with 92.1 lb (41.8 kg) wet for the previous generation, an insignificant difference actually based on the reduced amount of oil used. At 359.2 mm (14.1 in) long, it is 5.8 mm (0.2 in) shorter than the 2015 CVT. ■

*Paul Weissler*

# Nissan's Maxima has flowing sports car lines, upgraded 3.5-L V6 and CVT

**New Maxima features flowing lines, lowered center of gravity, and is slightly longer overall.**

A flagship car typically isn't the top seller, but it does set a tone for a product line. Nissan's 2016 Maxima, introduced at the 2015 New York Auto Show, put a strong exclamation point on styling, performance, and a full suite of electronic content in a four-door sports sedan. Maxima, now in its eighth generation, is one of Nissan's highest awareness vehicles, despite its comparatively modest annual sales of just over 50,000 units.

Maxima started in 1981 as a rear-wheel-drive car, but from the second generation (1985) it's been on a front-drive platform. In recent years, it had come a bit close to the Altima, but the new model creates needed separation, particularly in the styling area.

## Nissan cues plus sports sedan lines

The 2016 Maxima, which goes on sale this summer, has all the cues that say it's a new Nissan, plus the flowing lines of a sports sedan. The package called for a longer, lower car, with better fuel economy—so lighter weight and improved aerodynamics.

Although the new model is on the same 109.3-in (2776-mm) wheelbase, it's 2.2 in (56 mm) longer at 192.8 in (4897 mm). At 3471-3593 lb (1574-1630 kg), it's about 80 lb (36 kg) lighter than the previous edition. Torsional rigidity is claimed to be 25% higher, although Nissan provided no comparative numbers.

Under hood, the car gets a new version of the 3.5-L V6. Although the power number is up only modestly, from 290 to 300 hp (216 to 224 kW), the engine has more than 60% new parts including sodium-filled exhaust valves, which run cooler. Torque is 261 lb·ft (354 N·m).

There's a big improvement in fuel economy, up 15% to 22 mpg city/30 mpg highway, with some apparent assist from the CVT (continuously variable transmission). Nissan also has demonstrated that the Xtronic CVT can be an accepted transmission for a performance car. The ratio spread is wider at 6.3 (from 2.413: to 0.383:1) vs. the previous 5.4 (2.371: to 0.439:1). The new transmission provides stronger launch acceleration and the simulated shifting logic produces faster "shifts" (ratio changes) at large throttle openings.

Steering was changed from fully-hydraulic to electro-hydraulic.

**Maxima control stack has 8.0-in touch screen, which permits a reduction in number of buttons. A standard navigation screen can be "swept" over to the instrument cluster screen, which has a 7.0-in color display.**

For a smoother-running engine, the powertrain rests on six mounts, two vacuum-controlled, one liquid-filled, and three solid rubber, two of which are tied to torque struts. In addition, the oil pan was reshaped for greater rigidity to reduce engine vibration.

Although the Maxima has active noise cancellation to delete low-frequency sounds, it also has an Active Sound Enhancement system to provide a high-performance engine sound. If the driver engages the car's Sport drive mode, that sound hits an even more exuberant note. The drive mode selector also changes throttle response, transmission shifting, and steering effort.

A full range of safety electronics is a must in this market, and the new car has an available four-camera system (one in front, one in rear, one at each side) to provide a 360° view. Front and rear sonar sensors also are in the package.

## Predictive-warning system

The Maxima has a novel "predictive" collision-warning system based on a millimeter-wave radar sensor that looks beyond the car in front, to detect a possible obstacle (such as a stopped car) that could call for emergency braking. If this early-warning system detects the potential for a collision, in addition to driver alerts, it can brake the car to a stop.

All these sensors provide moving-object detection at speeds below 5 mph (8 km/h), blind spot warning, parking assist (from the sonar sensors) and intelligent cruise control.

The infotainment system includes a Drive Attention Alert that monitors driver steering patterns, a safety feature seen in the premium-car market. It issues warnings (including the coffee cup recommendation to stop) if it detects what appears to be driver drowsiness.

The instrument cluster has a standard 7.0-in color display and the center control stack incorporates an 8.0-in touch screen for NissanConnect with standard navigation and a built-in data modem for 911 notification as for airbag deployment. The stack, angled seven degrees toward the driver, is similar to what is used on the Murano, and it reduces the number of buttons from 25 to 10.

If the driver wishes, he/she can "swipe" the navigation screen from the control stack to the instrument cluster. A subscription-based Sirius Travel Link adds news, sports, fuel prices, etc. The console area adjacent to the cupholders has a Display Commander for control of the center stack display.

There are no options. The buyer can select from five trim/equipment levels: S, SV, SL, SR and Platinum, with the last two featuring LED projector headlamps and heated-cooled front seats.

The Maxima is assembled at Nissan's mammoth Smyrna, TN, plant, which has the highest output in North America (648,000 vehicles last year). The engines come from a nearby plant in Decherd, TN. ■

*Paul Weissler*

# Nissan arms its 2016 Titan with diesel V8, HD capability

**Unveiled at the 2015 NAIAS, the second-gen Titan is available in regular and XD models, with three cab types, two chassis, and three engines including a Cummins V8 diesel.**

Nissan officials admit it took a grueling 10-year development program to launch their second-generation Titan, but those waiting for the new truck simply groan and call it "late." Even as the company trickled out teaser information ahead of the 2016 model's debut at the 2015 North American International Auto Show, many among the Titan fan bases remained skeptical.

"Also in the news, Big Foot was spotted riding on the back of the Loch Ness monster," retorted one on Titantalk.com.

So when Renault-Nissan CEO Carlos Ghosn unveiled the new Titan at the Detroit show, he also was showing his company's determination to stay in the full-size pickup game. The original Titan, launched for the 2004 model year, was the Japanese OEM's first shot at breaking into the U.S. market's most lucrative and competitive segment. But the incumbent truck has been nearly invisible on U.S. roads outside of its production base in Canton, MS. To add to Nissan's frustration, development of the successor suffered false starts and delays, according to engineers involved with the program.

For their second-gen effort, Nissan is betting that bold product planning and clever engineering and systems integration will help turn the tide. For 2016 there are Titan and Titan XD models, the latter offering the compelling combination of an optional Cummins 5.0-L V8 turbodiesel and a robust chassis that stretches the gross-vehicle-weight rating to above 8500 lb (3855 kg). Presto, a light-duty platform that can deliver heavy-duty levels of towing (over 12,000 lb/5443 kg) and payload capacity (2000 lb/907 kg) with diesel muscle.

Peak torque of the Cummins V8 is SAE rated at 555 lb·ft (752 N·m) at 1600 rpm, and fuel economy is projected to be up to 20% greater than that of Nissan's current 5.6-L gasoline V8 (EPA ratings for the Cummins engine are still pending). To meet SAE J2807 towing standards, the Titan's cooling system capacity was increased more than 25% vs. the previous vehicle.

"The positioning is unique—we're the first to offer a V8 turbodiesel in a light-duty pickup, and in terms of capability we fit between the light- and heavy-duty territories," asserted Rich Miller, Nissan North America's Product Planning Director. "'Light-duty-plus' probably speaks best to what the new Titan really is," he said.

There are three cab configurations (single, crew, and king), three bed lengths, and five trim levels. For the first time Nissan will offer a gasoline V6 in addition to a gas V8 and the Cummins; all engines will be coupled to an Aisin AS69RC 6-speed automatic and American Axle drive axles. 4x4 models get a new transfer case, the supplier of which has not been announced.

The Titan program is North American to the core. Design/styling was led by Nissan's La Jolla, CA, design team under the direction of Diane Allen, who also headed the first Titan's styling. The Farmington Hills, MI, technical center led vehicle engineering, with nearly 50 staff from the original (A60) Titan team involved. The turbodiesel came

**Nissan North America's development team moved the new Titan's gearshift lever to the steering column, to open up center console space. Interior materials and switchgear on the pre-production trucks shown to media prior to the Detroit show were exceptional in their quality**

from Cummins' Indiana HQ and initial vehicle testing was handled at Nissan's Arizona proving ground.

Underpinning the Titan and Titan XD (tailgates wear an 'NP 1500' badge) is a modified F-Alpha platform that is shared with the Nissan NV commercial van and the Nissan Armada and Infiniti QX56 SUVs. The fully boxed steel ladder frame, supplied by Tower Automotive, is laterally and longitudinally stiffened for pickup duty and comes in two wheelbases: 151.6-in (3850-mm) for the XD, about 20-in (508-mm) longer than the non-XD Titan.

Other basic dimensions of the XD model available at the NAIAS debut: 242.9 in (6329 mm) overall length; 80.6 in (2047 mm) overall width; 78.7 in (1999 mm) overall height (4x4). Minimum ground clearance at the bottom of the rear axle housing is 9.3 in (236 mm). Claimed angle of approach is 21.4°; angle of departure is 22.8°.

Front suspension in both 2WD and 4WD models retains the upper-and-lower control arm layout with stabilizer bar; rear suspension is by leaf springs and twin-tube shocks. Tuning is entirely revamped for improved ride quality. "Rear coil springs were never part of the program," explained Gary Schoenborn, Vehicle Performance Manager at the Nissan Tech Center. "We feel we achieved excellent ride quality and hit our payload targets with the leaf springs."

The AAM rear axle features a 13-in (330-mm) ring gear and uses 3.5-in-diameter (89-mm) axle tubes—robust indeed for a light-duty application. An electronic locking rear diff is optional. The front differential on 4x4 Titans measures 9.25 in (235-mm). Titan's four-wheel-disc foundation brakes also could be mistaken for HD equipment, with 14.2 x 1.5-in (361-mm) front and 14 4 x 1.2-in (366-mm) rear ventilated rotors.

Titan's steering system remains the NV's hydraulic recirculating-ball type, which puts the new Nissan a generation behind the electric systems of Ford, GM, and Ram.

## Tumultuous beginnings

As one Nissan engineer observed to *Automotive Engineering*, "If the Great Recession hadn't happened, the Titan program might look a lot different. But we also learned about collaboration in the process." That's because Nissan had planned to partner with Chrysler to co-develop the second-gen Titan on the Ram 1500 platform scheduled for 2012—and assemble it at Chrysler's Saltillo, Mex., plant. That plan was understandably axed by new Chrysler owner Fiat.

The tumult ensured that a clay freeze on the second-gen Titan, and the truck's production timetable, would be delayed. But an upside emerged: Nissan hired Ram's hard-charging CEO, Fred Diaz, away from Chrysler in spring 2013, and Diaz is said to have quickly pushed through an iteration of the Cummins ISV5 0 V8. A diesel engine was not in the original product plan, and Diaz believed one would help differentiate Titan and ignite sales. Nissan and Renault diesels were initially discussed, but the Cummins brand resonates in the U.S. Fortuitously for Cummins its new 5.0-L V8, developed to serve both the light-duty pickup and commercial-vehicle segments, was itself struggling to find a customer.

Chrysler Powertrain had investigated the ISV5.0 for Ram use but determined it unlikely to meet fuel economy targets (above 25 mpg) levels, according to Ram Chief Engineer Mike Cairns. But Nissan and Cummins already had close ties: the ATLAS (Advanced Technology Light Automotive Systems) program. Launched in 2010 with the U.S. Dept. of Energy as a research partner, ATLAS aimed to

**Nissan's NASA-inspired "zero gravity" seats are standard in the 2016 Titan.**

develop an advanced 2.8-L four-cylinder diesel for use in future light trucks. Two prototype engines were installed in a pair of Titans to serve as testbeds for new technologies (including cooled EGR, close-coupled aftertreatment, and turbocharger innovations). According to Dave Goggin, Cummins' Pickup Markets Director, the ATLAS work paved the way for the V8 diesel in the 2016 Titan.

## Cummins' technologies for mpg, low emissions

The 2016 truck retains many customer-delight features of its predecessor—including rear doors that swing open 168° on Multimatic hinges, Utili-track flexible bed storage system, and spray-on factory bedliner. The cargo compartments integrated into the rear bodysides have been replaced by removable locking storage bins that fit within the bed over the wheel arches.

The new truck also adds hydraulic cab mounts on the XD model as part of an enhanced NVH-abatement package. "In our NVH benchmarking, the Chevrolet Silverado blew us away on interior quiet—they really stepped it up in that area. We'll be better," Schoenborn promised at a pre-NAIAS media backgrounder on the truck.

The thick underhood noise blanket on XD models points to the integration of the Cummins V8 into the Titan architecture. But this is a diesel designed to reduce NVH at its sources. The V8 Turbo Diesel (not to be confused with the ISV5.0 from which it was derived) features significant production firsts for Cummins. It marks the first use of a compacted-graphite iron cylinder block, paired with aluminum-alloy DOHC cylinder heads and composite camshaft covers. The V8, rated by SAE at 310 hp (231 kW), is also the first application of Cummins' new M2 two-stage turbo system with Rotary Turbine Control (RTC).

Developed by Holset, Cummins' U.K.-based turbocharger group, the M2 packages two turbochargers, one larger than the other and operating in series (not to be confused with a twin-turbo arrangement). The M2 system has four main operating modes: high-pressure two-stage; low-pressure single-stage; wastegate, and regen—the latter works to optimize operating conditions for the diesel-oxidation catalyst, SCR, and particulate filter (DPF).

Holset's new RTC is actually an electrically controlled rotary valve that directs the exhaust flow to either of the two turbochargers or to a wastegate depending on load. At low engine speeds the valve vectors 100% of the exhaust gas through the smaller turbine. As airflow increases with engine speed and load, the valve rotates toward the larger turbocharger (while allowing some gas flow to spin the small low-pressure turbo). The RTC can perform like a throttle to increase backpressure, thus serving as an exhaust brake and also raising exhaust temperatures to help to burn off ash that is collected in the DPF. It can also help to regulate boost pressure by routing excess gas around the turbines.

The V8 also features cooled EGR, a Bosch high-pressure common rail injection system and Bosch ceramic glow plugs.

## NASA "zero gravity" seats standard

In the all-new cab, the design team focused on improving ease-of-use and utility, as well as driver and passenger comfort. Front storage was increased 33% and rear storage 28%, the engineers said, compared with the previous truck. All switches are now grouped by function type and their location is based on frequency of use. The transmission shifter was moved to a lever on the steering column; this opened up considerable storage space in the center console that can now hold a 15-in laptop. The Titan's IP has a 5.0-in color display, unique diesel monitoring and maintenance displays, and 7.0-in NissanConnectSM with Navigation, Mobile Apps, and Services color monitor.

Titan also adopts as standard Nissan's "zero gravity" seats that are claimed to have been inspired by NASA. The seats, with optional heating and cooling, debuted on the 2013 Altima. The shape of their seatbacks and their patented structure provide varying levels of support in the lumbar, midback, and shoulder regions to position the occupant's spine in its natural position. For more on the seats, see http://articles.sae.org/11073/. Details of Nissan's early seat investigations can be found in these SAE Technical Papers: 2006-01-1302 and 2007-01-0348.

A comprehensive suite of electronic features help improve Titan's safety and dynamic performance with and without a trailer. They include Hill Descent Control, Hill Start Assist, and Brake Limited-Slip Differential (BLSD); Integrated Trailer Brake Controller, Trailer Sway Control (TSC), Tow/Haul Mode with Downhill Speed Control, and a Trailer Light Check system that allows hook-up by one person, plus checking turn signals, brake lights, and running/clearance lights from inside the cab.

The trailering package includes available RearView Monitor with Trailer Guides, and an Around View Monitor (AVM); the latter provides a bird's-eye view of the surrounding area from front, rear, and side cameras to help with parking and pulling in and out of tight spaces. The system also includes moving-object detection, which helps the driver detect moving objects such as vehicles, shopping carts, or other large objects when reversing via an on-screen notification and warning chime. ■

*Lindsay Brooke*

# New Astra is Opel-Vauxhall's "no excuses" car

**Hurrying to catch its design and engineering rivals: GM Europe's new Astra.**

General Motors' European duo, Opel and Vauxhall, have created a convincingly fresh design and engineering challenge to Volkswagen's ubiquitous Golf and Ford's Focus.

It's no mean feat. Although recent Astra models have been steadily gaining commercial and critical credibility, they generally haven't managed to attain the overall respect achieved by their archrivals.

But an all-new lightweight vehicle architecture and enhanced chassis and engine capability complemented by impressive added technologies and an interior that provides a surprisingly up-market ambience combine to give the new car added status.

Vauxhall's UK Ellesmere Port manufacturing facility is the lead plant for the new model (the other is Opel's Gliwice, Poland factory; design and engineering of the car is centered at Opel Germany) producing both Opel and Vauxhall badged cars in hatchback and wagon (Sports Tourer) configurations.

Depending on the version and its trim level, the latest generation Astra is up to 200 kg (440 lb) lighter than its predecessor. Body-in-white weight is down by about 20% to 280 kg (617 lb).

The new chassis is also lighter by 50 kg (110 lb) despite the fact that it uses no aluminum. Instead, Opel-Vauxhall incorporates high-strength (HS) and ultra-high-strength (HSS) low-weight steels and compact subframes as cost-effective solutions.

This has brought other pluses, said Horst Bormann, Opel-Vauxhall's Director of Vehicle Performance: "For example, if we reduce bodyshell and suspension weight we can have smaller brakes and that reduces weight further. We can also have smaller wheels and smaller tires—again weight saving."

To compensate for the aesthetic effects of wheel and tire changes, Design Director Mark Adams gave the

**Ellesmere Port in UK is the lead plant for Vauxhall/Opel Astra production.**

**GM Europe's weight saving campaign for the new Astra led to a 20% lighter body-in-white.**

Astra a lower, sleeker look, with a falling roof line to the distinctive C-pillar, which helps to enhance the car's individuality, providing " down road graphics" that can visually separate it from other vehicles in its class. Said Adams: "Basically we have done everything from scratch; the old car was too heavy and we needed to address that, so we fixed the whole package. This is a 'no excuses' car now."

Styling was inspired by Opel-Vauxhall's 2013 Monza concept.

Adams and his team, working closely with aerodynamicists, have managed to give the Astra a good if not exceptional Cd figure for its class of 0.285. Tweaks that have helped towards this include the placing of exterior mirrors on stalks, with benefits also coming from the C-pillar shaping.

Stefan Fesser, Plant Director at Ellesmere Port, where £140 million has been invested over the past 18 months, said the manufacturing process now includes laser brazing: "Opel has not used this before; we can work with HSS and no longer require spot welding of the roof, saving roof rails and bringing a huge weight reduction."

The Astra hatchback is slightly smaller externally than the outgoing model, by 5 cm (2.0 in) overall and 2.5 cm (1.0 in) in height, but despite a wheelbase shorter by 2 cm (0.8 in), interior space has increased thanks to the car's new architecture. Unlike the previous platform/architecture, which had to be shared with other distinctly different models such as the Zafira, the new car is using a dedicated solution just for the Astra range. The wagon (Sports Tourer) is expected to be joined by further variants.

At launch, the powertrain line-up for the new Astra range includes diesel units ranging from 74 to 148 kW (99 to 198 hp). Particularly interesting is the diesel 1.6-L CDTi 81-kW (109-hp) ecoFLEX, with stated 82-g/km CO2 emissions and combined fuel consumption of a remarkable 3.1 L/100 km.

A three-cylinder 1.0-L turbocharged gasoline engine is also offered, with 77 kW (103 hp) and maximum torque of 170 N·m (125 lb·ft) delivered at a very low (for a gasoline unit) 1800 rpm. CO2 emissions are stated at a best—for the ecoFLEX version with an Easytronic transmission, stop-start, and low rolling resistance tires—of 96 g/km (99 g/km with the standard five-speed manual), with combined fuel consumption of 4.2 L/100 km. The best zero to 97 km/h (60 mph) time is 10.5 s.

The company is slated to introduce 17 new engines by 2018.

Although Opel-Vauxhall does not pretend that the Astra is entering premium-market territory, its interior does have a perception of "premiumness" about it, particularly in the high-specification versions. Adams' description of it blending "sculptural dynamic shapes, premium materials, and an absolute focus on quality of execution and craftsmanship" may be a shade too lyrical, but it does indicate the essence of the cabin.

Another interesting aspect of the car's rather cheeky up-market technology message is the option of IntelliLux LED matrix headlights, a system until now firmly associated with the premium sector via Audi and supplier Hella. Opel-Vauxhall's supplier is the Austrian company ZKW. Each headlight comprises eight LED segments, the system automatically and constantly adapting the length and distribution of the main beam. IntelliLux works in concert with a front-mounted camera that identifies oncoming vehicles and switches off relevant LED segments to prevent dazzle.

As the vehicle exits an urban area, high beam lighting is automatically selected and remains on. Studies by the Technical University of Darmstadt, Germany, and the European LightSightSafety Initiative have confirmed the potential for increased safety offered by the system. This editor sampled its impressive efficacy during a night drive in the UK.

The Astra's technology repertoire also includes torque vectoring and electronic drag torque control to avoid wheel slip caused by high engine drag torque, helping to maintain vehicle steerability.

Suspension is fairly conventional with MacPherson struts front (decoupled to separate the paths of spring and damper settings) and a Watt's linkage at the rear. Opel-Vauxhall claims to be the first in the world to combine a torsion beam axle with the Watt's link—and patented it to prove the point. It has been further developed for the new Astra.

Chassis specialist Bormann regards it as providing the dynamic advantages of an independent multi-link setup without the penalties, such as added mass, complexity, and packaging. He added: "We have a new rigidly mounted suspension subframe, which is important. The old car had a much heavier subframe mounted with four bushings. We did a lot of structure test analysis and transfer path analysis and found the rigid mounting to be much more efficient with regard to mass, stiffness, and bending characteristics."

Torsional rigidity of the body structure is 19,600 N·m (14,400 lb·ft) per degree, a 400 N·m (295 lb·ft) per degree improvement over that of the previous Astra.

Front/rear weight distribution of the hatchback is 60/40 with the battery in the trunk.

The new car is the first Astra to feature OnStar, described by the company as "the ground-breaking connectivity technology now available across Europe." OnStar includes high-speed 4G LTE mobile network, emergency response notification, and stolen vehicle recovery alert.

**Exterior mirrors on stalks and subtle C-pillar shaping have helped lower the latest generation Opel-Vauxhall Astra's Cd figure.**

**High-tech LED matrix headlights are on the new Astra's options list.**

Astra is described by Opel-Vauxhall as being one of the first vehicles from a mainstream manufacturer to get Apple CarPlay and Google Android Auto, available through a new version of the IntelliLink infotainment system.

It all adds up to a potentially big step forward for GM Europe. ■

*Stuart Birch*

# Opel designs a new Astra

**Astra five-door is lighter and more compact than the outgoing model.**

Opel launched the fourth-generation Astra at the 2015 IAA (Frankfurt Motor Show). The car is more compact than the model it replaces and weighs up to 200 kg (440 lb) less. A five-door hatchback and five-door Sports Tourer estate were on display at the show.

The Astra (a Vauxhall in the UK) will be the first vehicle to feature General Motors' OnStar Connectivity system in Europe and comes with the option of matrix LED headlights, which Opel claims is the first time this feature has been offered on a European C-segment car.

"There was a lot of really close work between our engineering, product planning, marketing, and design guys on a daily basis," said Malcolm Ward, Opel/Vauxhall Exterior Design Director. The new car uses the Opel design language dubbed "Sculptural Artistry."

"That is really all about very fluid sculptural shapes with some very precise technical elements," explained Ward. "On a secondary level, we always try to give a lot of emphasis to the wheels, so we sculpture the surfaces very strongly so that the car sits better on the road. We are always trying to make our cars look as low and as wide as possible.

"When the car actually gets lower, then that's already helping us; that's the first point from the architectural standpoint," Ward continued. "But if you actually look at the car itself, just the way we have positioned a couple of the lines, you can see where the blade is, the undercut in the shoulder, but more importantly, how we have treated the whole C-pillar, what we call our breakthrough C-pillar. By stretching that black graphic as a window graphic, the whole window graphic still leans rearward, like in the previous generation, but with this breakthrough C-pillar, connects from the rear glass to the side glass and really makes the car look a lot lower than it actually is. That was the real technical challenge for our organization.

"Quite often manufacturers put a chrome accent on the belt," Ward added. "Again, that's a visual trick to try and get it to look lower. We deliberately put it on

**The Astra interior will feature OnStar for the first time in Europe.**

the DLO (daylight opening, above the doors) on the upper part, not on the lower part, because actually, with that breakthrough C-pillar, your eye automatically goes there. Then the chrome element gives a connecting link from the front to the rear that tells you that's how low the car is and in actual fact it's another 30-40 mm higher, as you can see. That was one of our big visual tricks."

The new Astra's track is wider than for the outgoing model, which also gives the design team more opportunity to work on the visuals of the car by exploiting the width around the wheels. Ward explained: "That gave us the chance to get even more sculpturing in the wheel arches, which again catches more light, and when it catches more light the car looks more efficient. When shadow and light play together in the right way, that's how you can make the car look like it's a lot more athletic and more efficient."

**Astra Sport Tourer made its debut at 2015 IAA.**

Any long-running series of models takes each new version through a change of profile, but Ward and the Opel design team wanted to retain the coupe-like profile of the outgoing Astra. "Only by staying with the overall profile of the car can you really communicate how much more efficient, how much lower and wider it has become. When you change the whole profile, a customer has no reference," said Ward.

This presents other challenges because aerodynamically, a more vertical rear is preferable. "You can see it's really pushed forward and pushed down, which is bad for aerodynamics," explained Ward. "So just by choosing this profile, we had lots of discussions at the beginning of the program. We convinced the organization that this was the right thing to do for this car, for the brand to really signal that even with a more difficult profile, it can be very aerodynamic. We committed to compensating for the lack of an upright rear in other parts of the car."

The sculptured and dynamic design theme is what the Opel design team has set out to bring to the interior too. "We've given the whole interior a very horizontal accent," said Ward, "At the same time we have simplified all the interfaces that you have with the car. For example all the HVAC buttons are linked together and separated from the main infotainment area. The infotainment system is quite Apple-like in its appearance with some fine chrome elements that give it this premium appeal. I think that's what this interior is all about." ■

*John Kendall*

# No 'shaky start' for Opel/ Vauxhall's new Corsa triple

**Opel Group's fourth generation Corsa takes the company's three-cylinder 1.0-L engine into high volume production.**

Thorsten Kniesa likes a quiet life, and as a senior member of the team that created Opel Group's (Opel and Vauxhall) three-cylinder Ecotec gasoline engine, he has helped ensure that buyers of the just launched Corsa hatchback can also have one.

If any compact car demonstrates convincingly the potential of a 1.0-L triple linked to a six-speed manual transmission as a downsizing powertrain solution, it is the little fourth-generation Corsa. And probably the most significant single mechanical contributor to that peaceful traveling is the use of a counter rotating balance shaft.

"It is a highlight for a three-cylinder, high-volume production engine, and it is there simply for reduced NVH (noise, vibration and harshness) reasons. Yes, it adds a little cost and weight, but it is a very good investment," said Kniesa, Assistant Chief Engineer for three-cylinder engines. "Without it, the engine would have different vibrations, and they would be felt by the occupants. This would be particularly noticeable during start-stop operation. Without it, there would be what we call a 'shaky start,' reminding the car's occupants that there is one cylinder less than normal. They don't want to know!"

The Opel Group (the nomenclature replaces GM Europe and includes all General Motors activities in Europe including Russia) triple, previously detailed by http://articles.sae.org/12970/, is also available in the smaller Opel/Vauxhall Adam, but with Corsa it is now entering the big time in volume terms.

Both engine and car were wholly developed in Germany at Opel's Rüsselsheim Technical Center. The engine follows Ford's highly successful and impressive three-cylinder EcoBoost, a power unit that has a distinctively sporty note when extended, a feature that many buyers savor. However, Kniesa says that such a sporty aural signature was not what Opel Group aimed to achieve; exceptionally low NVH was the target, and the balance shaft was part of that.

Ford says that the EcoBoost's flywheel has been deliberately unbalanced to deliver required levels of smooth running, without the necessity for a balance shaft, which would reduce engine efficiency. Volkswagen also does not use a shaft on its 1.0-L gasoline triple. The company claims that reciprocating parts are so light that a shaft is not necessary, although its triple 1.4-L diesel does have a shaft.

Andy Delicata, Ford of Europe Manager, Powertrain NVH, explained at the engine's announcement that the traditional way of reducing shaking forces in small-displacement engines is to install a counter-rotating balance shaft to cancel out most vibrations. But the problem with a balance shaft, he said, is that it is heavy, expensive, and it reduces fuel economy.

So his 1.0-L NVH engineering team attacked the problem by focusing on two areas: the engine's front pulley and rear flywheel, and the mounting system that connects the powertrain with the car's body. He said that the pulley and flywheel are unbalanced with weights that are placed precisely to counteract the natural shaking forces of the engine and drive the energy in a less sensitive direction. The engine mounts are designed to decouple as well as absorb the engine's shaking forces.

But Kniesa said Opel Group is confident that the benefits of a shaft overshadow the downside of its use. The shaft adds around 2% to engine unit weight—the dressed Ecotec unit has a mass of 106 kg (234 lb)—and up to 5% to unit cost, both figures with which Opel and Vauxhall are comfortable.

Significant though the balance shaft is, it is just part of an integrated assault on the new Corsa's NVH. "For the Ecotec program, there were three engineering tasks for a three-cylinder, continuously variable valve timing engine: best refinement, together with very good fuel efficiency and performance," said Kniesa.

How to benchmark refinement was an interesting challenge, he said, because 42 months ago when the engine and new Corsa programs started, there were very few triples in production: "At the time, three-cylinder engines were generally regarded as being an affordable entry-level for small cars, without the stress on refinement."

GM had a 40-kW (54-hp) non-turbo entry-level gasoline triple in 1997, which was not the most impressive power unit of its day.

"But now for the new Corsa, we are introducing our three-cylinder with a turbocharger, direct fuel injection–and the balance shaft. So we set our own standards and successfully compared it to our four-cylinder engines for both NVH and performance. Using direct injection and a high compression ratio (10.5:1), you get closer to the sound of a diesel. The high-pressure fuel system delivers 200-bar pressure, so we decoupled it from the cylinder head; there are no direct metal-to-metal components. We did the same for the heat shield, which can act as a sounding board; it was decoupled and stiffened."

But getting NVH down to really low levels called for far more, including great attention to fundamental design and a very stiff architecture for the all-alloy engine, bed plate, and two-piece oil pan: "All components had to be optimized for the required sound level, which means shielding, decoupling, and encapsulating."

The engine's behavior also contributes to the perception of quality and the reality of low NVH. Available initially at two power levels, 66 and 85 kW (89 and 114 hp), both versions achieve maximum torque of 166 N·m (122 lb·ft) at a low (for a gasoline engine) 1800 rpm, reducing the necessity to take it to (noisier) high revs to extract good performance. The six-speed gearbox through which it drives has been newly designed to minimize friction, is compact, and has a mass of 40 kg (88 lb), making it one of the lightest six-speeders on the market, claims Opel Group.

While the three-cylinder engine represents the main mechanical interest of the new Corsa, other engines include an improved 1.3-L CDTi (diesel) with CO2 emissions of 85 g/km and 1.2-L and 1.4-L gasoline units including a turbo but designed for economy rather than performance. The latest Corsa range also has a new chassis and sits 5 mm (0.2 in) lower than the outgoing model.

Virtually everything forward of the A-pillars is new or upgraded. "The chassis does not inherit a single component from the previous model," said Michael Abelson, GM's Vice President Vehicle Engineering in Europe. "We turned every conceivable setscrew to achieve a better drive experience with new levels of comfort, precision, and pleasure."

**The new Corsa's three-cylinder 1.0-L engine gives it 195-km/h performance.**

MacPherson struts at the front receive a new geometry and new knuckles to improve roll behavior and steering response. Rear suspension includes a compound crank with coil springs. The steering is electric with a City Mode for low-speed maneuvering. A new, reinforced subframe increases torsional rigidity.

GM used Millbrook Proving Ground in the UK as part of its chassis development program together with British public roads, notorious for often poor surfacing and, on minor roads, a plethora of potholes. A prototype was taken to Britain for several weeks of testing 12 months before the car was launched.

The new Corsa's curb weight is similar to that of the previous generation. The 85-kW triple three-door weighs from 1045 kg (2304 lb) plus 75 kg (165 lb) for the driver. The five-door model starts at 1088 kg (2399 lb) plus 75 kg for the driver depending on specification, about the same as the previous generation. Length is 4020 mm (158.3 in), width 1944 mm (76.5 in), and height 1479 mm (58.2 in).

The 85-kW Corsa reaches 100 km/h (62 mph) from rest in 10.4 s and has a 195-km/h (121-mph) top speed. The lower powered version achieves an official combined fuel consumption of 4.3 L/100 km and CO2 emissions of 100 g/km.

The Corsa's design was overseen by British Design Director Mark Adams, who has been responsible for giving Opel/Vauxhall models a distinctive brand signature including the "blade" running across the lower door sections.

Like Ford with its triple, Opel Group will further develop the Ecotec with more powerful options in prospect, possibly with an electric compressor configuration to improve turbocharger efficiency. The six-speed manual gearbox at present has a 30 N·m (22 lb·ft) margin before reaching its 200-N·m (148-lb·ft) limit so a significant increase in torque is also likely. ■

*Stuart Birch*

# Scion debuts iA and iM at New York

**The Mazda2-derived Scion iA is a front-drive subcompact without the expected understeer, according to Scion.**

Toyota's Scion youth brand has recently gotten ironically old. The company addressed that situation with a pair of new models at the 2015 New York International Auto Show that aims to refresh Scion's lineup and its image.

Scion's mission, according to Vice President Doug Murtha, is to attract new, younger customers to Toyota with distinctive products, to sell those products through a unique one-price process, and to give parent company Toyota a venue to experiment with new solutions.

The new Scion iA and iM provide some much-needed fresh product to the company's line, products that it hopes will attract non-Toyota customers to the brand.

One way Scion stated its youth credentials in the past was to eschew anything so boringly mainstream as a conventional three-box four-door sedan. Now it offers the iA, the product of an experimental joint venture with Mazda to provide Scion with a sport, affordable sedan built on the underpinnings of the Mazda2 and built in the NAFTA zone in Mazda's Salamanca, Mexico, plant.

Scion's goal is to attract young buyers with abundant content and the opportunity to drive a car that looks less like a starter car than the subcompact hatchbacks that are typical in its class. The iA is powered by Mazda's 106-hp (79-kW) 1.5-L four-cylinder and drives the front wheels through either a six-speed manual transaxle or a six-speed automatic.

Despite a no-haggle, single-specification approach, the approximately $16,000 iA will include 16-inch aluminum wheels, keyless entry, pushbutton start, tilting and telescoping steering, a collision warning system, and a rear back-up camera. There is even a standard 7-in dashboard multimedia display with voice recognition.

These are likely the kinds of features that will appeal to the target customers, but in case there are still shoppers who are more concerned with what's under the hood than in the dashboard, Murtha promises the Mazda-based underpinnings will deliver. "Just wait 'til you get behind the wheel and experience iA's surprising dynamic balance and lack of typical understeer," he boasted.

At the same time, if gas prices go back up, budget-minded drivers will appreciate the fact that their well-equipped, fun-to-drive, grown-up-looking car will also score 42 mpg on the EPA highway fuel economy cycle.

**The iM includes a standard Pioneer entertainment system, with a 7-in in-dash display.**

The iM, on the other hand, hews more closely to Scion's regular playbook, as it is a five-door hatchback heavily based on a Toyota model sold outside the U.S. In this case, it is the Toyota Auris that is rebadged and packaged as a Scion. An earlier version served as the base car for the Lexus CT200h hybrid.

The iM is built on the platform used by the current Corolla, but with a double-wishbone independent rear suspension in place of the Corolla's torsion beam axle for improved ride and handling.

This sub-$20,000 model is a step up from the iA in Scion's model line, with features like a 137-hp (102-kW) 1.8-L four-cylinder featuring Toyota's Valvematic continuously variable valve timing and lift system. The six-speed manual transmission includes hill-start assist so drivers new to shift-it-yourself technology don't embarrass themselves when they have to launch on a steep hill.

**The iM five-door hatchback sticks closer to Scion's original script of offering hatchbacks with more-than-expected equipment.**

The automatic transmission choice for the iM is a CVT that Scion says has been tuned for sporty response. To nudge the CVT further in that direction, there is a Sport mode that modifies the transmission's movement through its ratios as well as the car's electric power steering assist and the mapping of the engine's throttle-by-wire.

The CVT can also pose as a manual, with a manual mode that lets the driver click through seven pre-set ratios to simulate the experience of a conventional gearbox with a set of fixed ratios.

The included six-speaker Pioneer sound system uses an in-dash head unit that has a 7-in display screen with HD terrestrial radio and the Aha Internet radio app. Its air conditioning is the dual-zone variety, just like in luxury cars, its heated outside mirrors are power folding, for urban parking security, and its headlights turn on and off automatically.

Both cars will arrive at U.S. Scion dealers in September. ■

*Dan Carney*

# Smart collaboration results in new Fortwo

**New Smart Fortwo has some familiar styling cues, but the engineering package is new, developed by Mercedes-Benz with Renault.**

The third-generation Smart Fortwo has a new look, but with some familiar styling cues of its predecessor, and more importantly it incorporates significant changes under the skin. Overall length remains the same at 106.1 in (2694 mm), and the 73.7-in (1872-mm) wheelbase sees a modest 0.2-in (5-mm) increase. The car has been on sale in Europe for five months and just made its U.S. debut at the 2015 New York International Auto Show.

The new model was co-developed by Mercedes-Benz and Renault, and many of the parts, including the basics of the platform and the powertrain, reportedly are shared with the somewhat larger Renault Twingo.

## Turbocharged three-cylinder

The rear-mounted, rear-drive powertrain represents what will be the greatest difference in the driving experience. The car goes on sale with a 0.9-L three-cylinder, turbocharged with an electronically controlled wastegate, and rated at 90 hp (67 kW) and 100 lb·ft (136 N·m). This engine will represent a major power increase over the 70-hp (52-kW) naturally aspirated 1.0-L three-cylinder previously used. However, a lower-output 71-hp (53-kW) engine also is available in Europe and reportedly will come to the U.S. later.

The new transmission offerings are significant. The previous automated manual transmission was much maligned for jerky shifting. There's now a five-speed manual and a six-speed DCT (dual clutch transmission). Although DCTs, particularly lower-end versions with dry clutch packs, are not noted for their smoothness, this one is likely to provide significantly better shift quality than the previous automated manual.

The new model adds a third "door," actually a horizontally-split tailgate (top half goes up, bottom half goes down) to permit easier loading of the cargo area, particularly in tight parking spaces. The passenger's seat also easily folds flat to increase capacity. Passenger compartment width has increased by 104 mm (4.1 in) to 1663 mm (65.5 in) to provide more comfortable seating positions.

Operation in the tight quarters of city driving has always been a Smart's hallmark, and the new model claims a record for smallest turn circle diameter (curb to curb) of 22.8 ft (6.95 m).

Ride comfort is always a challenge with such a short wheelbase, but engineering steps were taken to make improvements. The front MacPherson strut was redesigned and incorporates greater spring travel, for lowered impact over uneven road surfaces. In the rear the reworked DeDion axle separates the twin-tube shocks from barrel-shaped helical coil springs, which include elastomer spacers to reduce spring noise transmission to the body and rear axle. A Crosswind Assist system, which comes into play at 50 mph (80 km/h), uses the electronic stability control system's anti-lock brakes to steady vehicle tracking in windy driving conditions.

A Sport suspension with less coil spring travel is on the option list, but with the typical U.S. buyer more concerned about less bouncing on public roads, the take rate might not be significant.

## Safety upgrades

Although the previous Fortwo's high-strength-steel Tridion structure has delivered good ratings in U.S. National Highway Traffic Safety Administration-mandated crash testing, it saw less success in narrow-offset crash tests, such as the one performed by the Insurance Institute for Highway Safety. For the new model, the car was subjected to such tests against C-Class and S-Class cars and apparently performed well, as indicated in a Mercedes-Benz video that showed the Fortwo driver area was basically intact and the door could be opened.

The new Tridion cell had been significantly strengthened with greater use of ultra-high-strength steels, and Mercedes-Benz said it functions as a safety cage in severe collisions. Eight airbags are standard. The new Fortwo does add some weight as a result, with an increase of 132 lb (60 kg) to 1940 lb (880 kg).

Available safety options are Lane Keep Assist and Forward Collision Warning. The collision warning system incorporates mid-range radar sensing, with a warning light that goes on if the car is getting too close to one ahead. The system issues an audible warning if the electronics sense a danger of collision.

There are three trim levels: Passion (basic, but includes power-operated and heated mirrors), Prime (adds rain and light sensors, heated seats for driver and passenger, fog lamps, and panoramic roof), and Proxy (which upgrades the interior trim and adds stowage in the tailgate).

The Fortwo uses smartphone connectivity to a control stack touch-screen with integrated real time navigation. A free app is Cross Connect, which provides guidebook-like assistance and a feature called Car Info, which computes g force, fuel efficiency, and posts geo-identified access to Smart-size parking; plus an assortment of music options.

A new four-passenger model (Forfour) also has been developed; and diesel, electric vehicle, and cabriolet have been available in the past, so the 2016 Fortwo just introduced may soon have U.S. stablemates. ■

*Paul Weissler*

# Can Toyota's Mirai be the next Prius?

**Toyota thinks that its new Mirai fuel-cell car can open up new horizons like its Prius hybrid did.**

A decade and a half ago, low gasoline prices and rebounding car sales did nothing to stop Toyota from introducing its fuel-saving Prius hybrid-electric vehicle. During those years the Prius, whose name means "to go before," pioneered an entirely new green-car category while it burnished the company's environmental credentials. Hybrid-electric propulsion has now gone fully mainstream, with total U.S. sales of hybrids topping 3 million in late 2013.

Today, with fuel prices sagging and auto sales rising, the Japanese carmaker is again attempting to open up a whole new green-car segment with its first hydrogen fuel-cell production vehicle, the Mirai. Whether the Mirai, which means "future" in Japanese, presages Prius-like success is hotly debated.

## 'Like any other car'

"The Mirai is like any other car," Takeshi Uchiyamada, Toyota's Chairman of the Board and "father of the Prius," told the press who had assembled Nov. 16 in Newport Beach, CA, at a lead-in event for the Los Angeles Auto Show. "It has a 300-mile range, fills up in 5 minutes and accelerates from 0 to 60 mph in about 9 seconds," he said. "The big difference is that this car uses no gasoline and emits only water vapor from the tailpipe."

In the U.S., the four-door Mirai sedan will have a sticker price of $57,500. But with about $13,000 in federal and California state incentives and subsidies, that figure drops to $45,500. Toyota expects the majority of customers to lease the car for $499 per month for 36 months, however.

The initial production goals for the Mirai will be 700 units, according to the company. The car will be introduced into the Japanese market before year's end and will go on sale in Europe early next year. California dealerships will get the new car by summer 2015 and those in five northeastern U.S. states will receive the first shipments in 2016.

The Mirai, which is the production version of the FCV Concept that was unveiled at last year's Tokyo Motor Show, is built around a 114-kW fuel-cell stack that converts compressed hydrogen gas into electricity. That electricity is stored in a nickel-metal-hydride battery pack that powers electric motors, which drive the front wheels.

The car's proton-exchange-membrane fuel-cell stack, which is encased in a protective carbon-fiber composite housing, has a power density of 3.1 kW/L. Unlike many previous units, it requires no humidifier to keep the polymer membranes damp to promote proton transport. Instead, the system circulates the water that the stack itself produces. The compressor and blower whirs and whines that have accompanied most earlier fuel-cell systems have been mostly eliminated to help keep the passenger cabin quiet.

The Mirai contains a pair of 10,000-psi hydrogen storage tanks, one under the rear seats and another just behind them. The tanks, which have 62.4-L (16.5 gal) and 60-L (15.9 gal) capacities, respectively, feature outer protective shells composed of filament-wound fiber/resin composites that are lined with impermeable membranes.

## Why hydrogen?

"Why hydrogen?" Uchiyamada asked the crowd rhetorically. The rate of cost reduction for fuel-cell systems is falling faster than that of electrochemical batteries, he answered, noting that "a 300-mile battery would be much larger and heavier than a fuel-cell stack and fuel tanks. Our fuel-cell system is really just a better battery."

Uchiyamada added that the fuel-cell tanks contain enough hydrogen to power an average American home for a week in emergencies, should buyers choose to add an optional power take-off port.

At 1850 kg (4100 lb), the car is still 100 to 150 kg (220 to 330 lb) lighter than a Camry hybrid, said Satoshi Ogiso, Managing Officer at Toyota and lead on the product planning group for the Mirai. Engineers placed all the fuel-cell system components low in the frame to create a low center of gravity for good handling performance. The underbody of the car is completely covered to cut road noise and aerodynamic drag because the fuel cell's heat output is less than that of a traditional car engine.

The Mirai's powertrain is 95% cheaper to build than the fuel-cell system in the 2008 Highlander fuel-cell SUV, he said. Some of cost savings derive from the decision to adopt much of its drive system—controls, energy management system, and batteries—from the company's hybrid models. Beyond the use of common parts, a brand-new converter triples the system voltage, which saves space, weight, and cost. Other savings stem from minimizing the use of high-priced platinum-family catalysts in the stack, said Ogiso, who added that R&D work is continuing to develop less costly catalyst alternatives.

He explained that the Toyota fuel-cell system is fully scalable and that company planners are contemplating hydrogen-fuel vehicles ranging in size from forklifts to sports cars to trucks and buses.

Ogiso stated that the car's exterior was styled to be both "distinctive" and "futuristic," with a feeling of flowing water. Whether Toyota's design effort was successful depends upon one's taste, but it must said that the car has something of the look of a Cylon Centurion from "Battlestar Galactica" from the

**Most of the fuel-cell system components are placed low in the frame to improve the Mirai's handling.**

front. The huge air intake grilles are there not only help cool the fuel-cell system's three radiators, but feed the vehicle's air-management system. The rear end reminds many of the back of 1970s-vintage BMWs. Whatever the styling, the new Mirai has a slippery aerodynamic drag coefficient of 0.293.

## Building a fueling infrastructure

Although the company president and CEO Akio Toyoda in a video touted the Mirai as "a turning point in automotive history" that would cut dependence on oil and environmental degradation, fuel-cell vehicles face an impediment that hybrids do not: a dearth of hydrogen fueling stations and the costly supply infrastructure to support them. This classic "chicken-and-egg" dilemma accounts for much of the skepticism now aimed at the technology.

Opposition to the adoption of hydrogen as a fuel has arisen in the U.S. because much of today's hydrogen is made by steam reforming natural gas, which undermines its positive environmental impact. Tesla chief Elon Musk has, for example, called fuel cells "a load of rubbish."

To help address the fundamental fueling issue, Toyota is partnering with Air Liquide, a global producer of hydrogen and other industrial gases, to develop new regional markets, said Chris Hostetter, Group Vice President for Strategic Planning. Part of the initial plan is to install a dozen hydrogen stations across Connecticut, Massachusetts, New Jersey, New York, and Rhode Island by 2016. Using the Street, a computer model developed at the University of California - Irvine that estimates the number of hydrogen stations needed to supply a fuel-cell fleet, planners have selected strategic locations to create a "hydrogen highway" between the New York and Boston metropolitan areas, said Ole Hofelmann, CEO of Air Liquide.

Hofelmann said that the hydrogen should sell for around $10/kg—energy-wise, a kilogram of hydrogen is roughly equivalent to a gallon of gasoline. Because fuel-cell powerplants offer double the energy efficiency of conventional car engines, $10/kg translates into an effective price of about $5/kg, which would be more competitive with conventional fuels.

Toyota already has invested in hydrogen fueling infrastructure in California, which hopes to set up a hydrogen highway along the West Coast all the way up to British Columbia. UC Irvine's Street model, which keeps potential customers

**The Mirai's passenger compartment is reasonably roomy and comfortable.**

at all locations within 6 min of a station, indicates that only 65 stations can support some 10,000 fuel-cell vehicles in California. In May, Toyota announced a $7.3 million loan to FirstElement Fuels to support the operation and maintenance of 19 hydrogen fueling stations across the Golden State, which hopes to have 20 stations in operation by 2015, and 40 stations by 2016.

The tide is also rising for hydrogen fueling infrastructure elsewhere in the world: Germany now has 15 stations and has plans for 50 by the start of 2015 and 1000 by 2020. Japan currently has 17 stations, but hopes to have 100 by 2016. Korea plans to have 160 stations by 2020; the U.K. expects to have 15 stations by 2015 and 65 by 2020; and Denmark wants to install 15 stations by 2020.

Meanwhile, other automakers are launching hydrogen fuel-cell initiatives as well. Hyundai began selling a hydrogen-powered version of its Tucson SUV in certain California markets this summer. And at the LA Auto Show, Honda is displaying its latest, the FCEV Concept, which is to hit the market in 2016. The FCEV Concept builds on the company's six-year-old FCX Clarity leasing program. It has a newly designed 100-kW fuel-cell stack that, like Toyota's, features a power density of 3.1 kW/L. The fuel-cell system is one-third smaller than that in the FCX Clarity and therefore can fit entirely under the hood, Honda said. A single 10,000-psi hydrogen storage tank gives the FCEV Concept a range of 435 mi (700 km).

What perhaps distinguishes the Mirai from the others is that Toyota plans to add the car to its core product lineup rather than simply lease it to individuals. Toyota seems deadly serious about developing a hydrogen-powered "people's car," but Uchiyamada stressed that consumer education will be key to any eventual success. Whether fuel-cell vehicles are finally ready for the mass market "rests less on the car, and more on the ownership experience," he said. After all, it took a decade or so for the Toyota's global sales of hybrids to reach a million units, and today the Mirai faces a similarly long journey.

(To view a YouTube posting of the audio and Powerpoint slides from a recent U.S. Department of Energy Webcast reviewing SAE International standards on hydrogen refueling, go to http://energy.gov/eere/fuelcells/webinar-introduction-sae-hydrogen-fueling-standardization.) ■

*Steven Ashley*

# Toyota gives a look at its re-engineered 2016 Prius

**Toyota designed the rear end of the 2016 Prius with "unique and striking rear combination lamps" that has the side benefit of helping to differentiate the vehicle from previous generations.**

While Toyota threw quite the launch party in early September in Las Vegas—a first for the company—to enable over 350 guests to get an official look at the 2016 fourth-generation Prius hybrid as opposed to subsisting on leaked images and suspect drawings, the party was most definitely more about looks than details.

One detail that did become clear was that even if over the last 15-20 years the Prius has accounted for more than 3.5 million in global sales (a good chunk of Toyota's overall 8 million hybrid units), Toyota now seems to want the Prius to be viewed as just another pretty face, albeit one that happens to be a hybrid, hoping that the vehicle's design will tempt new buyers to enter the hybrid market. It is already among the top three of Toyota vehicles in terms of loyalty rates.

If Toyota says so itself, the 2016 Prius features a "striking exterior design [that] ushers in [a] new hybrid era." Besides the look of its redesigned exterior, emphasis has also been placed on performance with a "new platform and rear double wishbone suspension [that] dial up driving dynamics." As Bill Fay, Group Vice President and General Manager, Toyota Division said at the event, "What was once a rational purchase that for many customers focused on fuel economy, is now so much more."

**Toyota says the Prius' all-new rear double wishbone suspension produces a better connection to the road for greater control and feedback in all driving conditions. A more rigid body helps improve responsiveness while cornering.**

Toyota says its rationale behind the "completely reimagined" Prius' exterior was inspired by a runner in the starting blocks, with a "sporty design that conveys a feeling of forward motion" that it hopes will having future customers running toward it. Part of that design consisted of the vehicle's hood being made lower and the roof peak moved forward. Overall, the new Prius is 2.4 in (61 mm) longer, 0.6 in (15 mm) wider, and 0.8 in (20 mm) lower with more interior space than the model it replaces.

New standard LED headlamps were placed to minimize the Prius' front fascia and "help define the vehicle's character." The rear combination lamps are certainly "unique" as Toyota describes and follow along the lines from the rear spoiler to the trailing edges of the sides.

In terms of the details about some of the technologies that make up the Prius, including which battery options (nickel-metal-hydride vs. Li-ion) will be available for which vehicles, more detail on Toyota's New Global Architecture (TNGA), and confirmation or denial about an all-wheel drive version, Bob Carter, Senior Vice President—Automotive Operations, said that Toyota is holding onto that information and more until the Tokyo Motor Show in late October, when the company will "give you all the specs you can stand."

**Prius' wide instrument panel was designed to be both functional and provide "a sense of comfort and spaciousness."**

**The 2016 Prius will be available in seven exterior colors, including the new "Hypersonic Red." This new color uses a newly developed treatment process and "highlights the external forms."**

**Toyota designers included "bold character lines" that run along the side of the Prius from the front fenders, "one just above the sill that sweeps up toward the rear fender, while the upper crease morphs into the rear spoiler, which has been lowered."**

**Owner feedback motivated Toyota to design a new interior that "adopts advanced technology" and offers functionality and has a "high visual impact."**

For now, Toyota teases that the vehicle will have smaller, lighter hybrid system components; higher energy density in the batteries; and an internal combustion engine (probably 1.8 L) touting "ground-breaking thermal efficiency" (more than 40%), that will all contribute to about a 10% fuel economy improvement. Hinted at the Prius reveal, Toyota says an Eco model, to be unveiled at some other future date, will achieve an even greater improvement.

The 2016 Prius is the first global vehicle to implement TNGA, which is an integrated development program for powertrain components and vehicle platforms. Vehicle development is strategically grouped to promote better sharing of components to improve resource efficiency by over 20%. Just a sampling of the benefits the Prius reaps via the use of the TNGA is a lower center of gravity and a more responsive suspension package that includes the new rear double wishbone design.

TNGA is also expected to provide a more rigid structural framework to help enhance occupant protection with an upper body designed to distribute frontal collision impact energy.

The vehicle is expected to be in dealerships in early 2016. ■

*Jean L. Broge*

# 2016 Toyota Tacoma debuts at 2015 NAIAS

**The 2016 Tacoma is inspired by a sleek desert race track, according to Toyota's Bill Fay. "The Tacoma is all about recreational action and athletic agility," said Fay. Four different wheel designs, ranging from 16-in to 18-in, will be offered on the 2016 Tacoma.**

It's been 10 years since an all-new Toyota Tacoma launched in the North American market.

"The 2016 Tacoma has a new Atkinson cycle V6 engine, two new transmissions, a new transfer case, a new rear differential, and a platform change," Mike Sweers, Tacoma's Chief Engineer said in an *Automotive Engineering* interview at the 2015 NAIAS.

Simply put: "This is a big re-do," Sweers said.

Tacoma will again offer a four-cylinder engine, while a 3.5-L Atkinson cycle V6 with the automaker's D-4S technology featuring both direct- and port-fuel injection that replaces the current Tacoma's 4.0-L V6 engine. The new V6 can be mated to a new six-speed manual or a new six-speed automatic transmission. "In both transmissions we changed the bearing surfaces so that we have more efficiency," Sweers said.

While the current Tacoma's body shell is made of mild steel, the 2016 Tacoma uses high strength steel (HSS) and ultra HSS. "We have HSS through the whole body to reduce mass, and ultra HSS is in the side outer, A-pillar area, and roof reinforcements," Sweers said. "We've increased our yield strength substantially." The ultra HSS will be hot stamped in North America, a first for Toyota in North America.

The 2016 Tacoma frame uses HSS in the 440 to 590 range as a way to increase rigidity and overall strength.

Tacoma will be sold in five different grades, each with a unique personality and look. More than 40% of the Tacoma trucks sold are TRD models. The new Tacoma TRD Off-Road will feature a power outlet in the composite truck bed. And like all Tacoma models, a GoPro camera mount—located near the rearview mirror—is standard fare.

The Tacoma TRD also will offer the midsize segment's first Multi-Terrain Select. Five different settings help address throttle position and braking as a driver-assist for tackling mud, rocks, sand, and other types of terrain.

"We're also introducing Crawl Control, a technology that provides fully automatic acceleration and braking. All you do is steer. So if you want to climb a rock face or go down a rock face, the truck takes care of all that by itself. You can set the speed between 1 and 5 mph, and it will handle the braking and acceleration," Sweers said. "The weekend warrior will look like a pro out there."

The new Tacoma was designed and engineered in Michigan. Said Bill Fay, Toyota Group Vice President and General Manager, "It was tested and tuned at our proving grounds in Arizona and will be manufactured in our future home state of Texas as well as our plant in Baja California (Mexico)."

Tacoma has a more than 60% market share in its segment. "Even at the end of the current generation's 10-year lifecycle, it manages to outsell its closest competitor 2:1," said Fay. ■

*Kami Buchholz*

# Toyota chief talks trucks

**Aerodynamics was a major focus in the development of the 2016 Toyota Tacoma, and it will become increasingly important for trucks in the future, according to Mike Sweers, Chief Engineer of the Tacoma and Tundra pickup trucks.**

It's been 10 years since the last all-new Toyota Tacoma, which isn't unusual for a truck's life cycle, according to Mike Sweers (SAE Member, 2002), Chief Engineer for the Tundra (since 2010) and Tacoma (since 2012) programs at Toyota Technical Center in Ann Arbor, MI. Trucks require "a lot more resources to develop" because of product variability: "If we take the 2016 Tacoma, we have an Access Cab and Double Cab, we've got two bed lengths [60.5 and 73.7 in], two wheelbases [127.4 and 140.6 in], and from a powertrain standpoint we developed a new [3.5-L direct injection Atkinson cycle V6] engine and two new transmissions [six-speed automatic and manual], so all that stuff takes time," he said, adding that Toyota's emphasis on QDR (quality, durability, and reliability) also contributes to development timing.

The new V6 with VVT-iW (variable valve timing with intelligent wider intake) is equipped with Toyota's D-4S technology, featuring both direct and port fuel injection. The engine generates 278 hp (207 kW), an increase of 42 hp (31 kW) over the previous V6, and 265 lb·ft (359 N·m) of torque at 4600 rpm. Estimated city/highway/combined mpg for the 4x2 automatic is 19/24/21.

A few major achievements for the third-generation Tacoma program include a 12% overall drag reduction—"the largest reduction in Toyota truck history," a 43% improvement in NVH, and a V6 tow package for towing up to 6800 lb (3085 kg) per SAE J2807—a 300-lb (136-kg) increase over the previous V6. Sweers spoke with *Automotive Engineering* at a recent Tacoma launch event about the compact-truck program and truck trends in general.

## Are there any plans to bring a diesel to the marketplace?

We don't talk about future powertrains or development of future powertrains. The diesel question is interesting in this segment compared to the full-size segment. In full-size there's a big demand for diesel, but in this [Tacoma's] segment it really comes down to cost vs. return on investment. The biggest difficulty with a diesel right now is meeting [EPA] Tier 3 or [California's] LEV III emissions. As we move toward 2017 and 2020, there's a break point where the emissions requirements increase. Can it be accomplished?

**The Tacoma TRD Off-Road features a Multi-Terrain Select system (automatic transmission only) that allows the driver to choose between different types of terrain such as loose rock, mud, or sand. Each input regulates wheel spin by adjusting throttle and brake pressure to provide maximum traction.**

Yeah, but your aftertreatment system continues to grow in cost. So if you have an engine cost plus $3000 or $4000 for an aftertreatment system, and fluctuating fuel price, does the customer see a return on that investment? That's the struggle right now. On a full-size you can get it because there's a lot of residual value that goes with a truck, but we haven't seen a big demand in the compact segment for a diesel at this time. And we don't know if there's a return on that investment. It's a very interesting topic but the real question is, is it a one-year or two-year engine, or can it make it to 2020 when we have to be at ULEV70? That's where it becomes really difficult.

## Ford's lightweighting strategy for the F-150 is well known. Was aluminum considered for this program, or for Tundra?

You're always looking for different ways to skin a cat, and Ford picked aluminum. It's just like what you're seeing in our 2016 Tacoma, that whole cabin structure is all-new development, and their [Ford's] whole cabin structure was all-new development. Anytime you get a chance to start over and do something different, from an engineering standpoint, what a wonderful program to work on. But there's several ways to take weight out, and we tend to be a little more conservative. We use aluminum in different areas, but as you look at the ways to do it and the effect on cost of ownership for the customer, we chose high-strength and ultra-high-strength steel over aluminum [for the structure] because there's more repair shops that can handle that; everybody can work with steel. And the cost of ownership goes down because your insurance costs don't go up. In the future, you may see composites or carbon fiber or aluminum—there's several ways to get there. One of our competitors is working with magnesium sheets [General Motors announced in 2012 it was testing a thermal-forming process and proprietary corrosion-resistance treatment for magnesium sheet metal], so everybody's looking for what is the next breakthrough in body development. I think aluminum has its place on certain things, it's just that we have taken a different approach.

## What are some of the major technology trends moving forward in the truck market?

As we look toward the future come 2025 with CAFE and emissions regulations, I'm not sure what a truck is going to look like. For a full-size truck, the number one reason to buy it is capability or towing. For a compact truck, it's really about lifestyle and what that person does with their vehicles. So for us, our customers are the youngest and most active in the segment. What they're expecting as we move to the future, is it going to change? I don't think their requirements are going to change, but how we address CAFE and emissions is going to affect what that vehicle looks like. Breakthroughs in aerodynamics—we spent a lot of time on aerodynamics on this truck [Tacoma]—and how do you improve the aerodynamic efficiency of the vehicle without giving up its capability. That's going to be a big challenge. NVH is going to be a big challenge. How do we keep refining what the customer's seeing? What the customer would accept 10 years ago is no longer acceptable. What used to be considered luxury-class NVH is now down in our C [segment] cars and our mid- and full-size cars. So from a truck standpoint, the guy who came out of an SUV or a luxury car and decides he's going to be a truck owner now, he's expecting the same creature comforts, the same NVH treatments in this truck. And we're witnessing this in the explosion of the premium truck models now. So how do we get to that next level? Again from an engineering standpoint, what a great challenge.

## The Tacoma has a standard GoPro mount and Qi wireless charging. Is consumer electronics integration a big trend moving forward?

It depends on the segment itself and the demographics of the segment. In the compact-truck segment, these customers are tied into the latest electronics...For a Tacoma—that Qi wireless charging or that GoPro mount, those were ideas that came right from our customer base—'At home I throw my phone on a Qi pad but in the vehicle I've got to connect a cord and plug it in; if somebody else wants to plug their phone in, do I have enough USB ports or power points for them?' So listening to the voice of the customer, we're trying to apply what they're asking for.

We teamed up with GoPro in San Francisco, and it was a great development because our body engineers learned a lot—camera positions and how to make the mount correctly—and GoPro engineers learned a lot, because they're not familiar with working with the auto industry; this is the first time it's ever been done. We worked with their video team and mounts engineering team, and shared information and data back and forth to find the best location that meets all regulations, which they're not familiar with, but also how to get the best camera angle for our customers. Simple things like getting a perspective of the vehicle, which is something we learned from the GoPro engineers. When you mount that camera in our vehicle, it's capturing some of the hood so you don't just watch the ground and sky bounce around. The other thing is people like to see themselves and get their passengers' reactions, so you can flip the camera around and capture everybody in the vehicle but still capture some of the outside terrain so you know what's going on. ■

*Ryan Gehm*

# Volkswagen re-engineers Tiguan

**Forward-looking radar is standard equipment on the new Tiguan, promising the ability to automatically stop if a pedestrian is sighted ahead of the vehicle.**

Volkswagen's original Tiguan compact SUV of 2007 was too expensive and too thirsty to achieve blockbuster sales during the segment's emergence as a key source of sales. But it achieved worldwide sales of 2.64 million during its production run, providing a strong foundation for the next-generation version, introduced at the 2015 Frankfurt Motor Show.

The company has aggressively addressed the original Tiguan's shortcomings, building the new generation version on its high-volume, low-cost modular transverse matrix (MQB) platform used for the Golf and other compact models. This has produced a vehicle that is longer, lower, wider, and more car-like than its predecessor.

Illustrating the anticipated range of uses for a crossover SUV, the Tiguan debuted in both on-road and off-road-centric configurations along with a hybrid-electric concept of a future efficiency-focused Tiguan GTE model.

While the Tiguan's cargo capacity has increased 10% to 18.4 ft$^3$ (521 L), the vehicle is 110 lb (243 lb) lighter than before. Overall length is up 2.4 in (61 mm) to 176.6 in (4486 mm) and width is increased 1.2 in (31 mm) to 72.4 in (1839 mm). Height is reduced by 1.3 in (33 mm) to 64.3 in (1633 mm). The lower roof combines with other aerodynamic improvements such as wind tunnel-designed side-view mirrors to slash drag by 40% to 0.31 coefficient of drag.

Worldwide, Tiguan features a suite of eight different engines, all of which include turbocharging, direct injection, automatic stop/start, and battery regeneration as standard equipment. Half the engines are gasoline and half are diesel, with the gasoline family topping out with a 220-hp (164-kW) 2.0-L four cylinder and the diesel family headed by a twin-turbo 240-hp (179-kW) 2.0-L four. Overall, the engine families are 24% more efficient than those on the previous model, which were Euro-5 compliant and 10% more efficient than those that were Euro-6 certified.

The 4Motion all-wheel-drive system is optional with some engines and standard with others. Its new Active Control switch lets drivers set the parameters of its operation according to the conditions; on-road, off-road, and snow.

In normal on-road driving the Tiguan's Haldex coupling disconnects the rear wheels from the engine for maximum efficiency.

While cost containment and improved efficiency were critical upgrades to the Tiguan, Volkswagen has made important electronic safety aid systems standard equipment, making the Tiguan even more appealing to young families.

That means an automatic post-collision braking system and a forward radar system that watches for potential collisions and acts to prevent them. Both are standard on all new Tiguans. If the radar spots a person in the road ahead, it issues an alert followed by automatic braking if the driver doesn't act. Automatic lane-keeping assist that steers the Tiguan back into its lane is also standard.

Tiguan's crash-safety extends to pedestrians, with the use of a pyrotechnic hood release that pops the hood open in the event the car strikes a person, providing additional crush space between the sheetmetal and the engine beneath.

While the new Tiguan goes on sale in 2016, the company is less specific about plans for the hybrid-electric concept version also shown at Frankfurt. Its internal combustion engine is a 115-kW (154-hp) turbocharged gasoline four-cylinder, which combines with an electric motor for a maximum system output of 160 kW. They drive the vehicle through a six-speed dual-clutch transmission.

The 13 kW·h lithium-ion battery pack can propel the Tiguan GTE concept 31 mi (50 km) on electric power alone, with an electric top speed of 130 km/h (81 mph). Using the electric motor to boost the gasoline engine's performance produces an available top speed of 200 km/h (124 mph) and 0-100 km/h (0-62 mph) acceleration of 8.1 s.

VW also imagines the possibility of a roof-mounted solar panel that can, over extended periods of time, provide a worthwhile amount of free energy. Though a solar panel cannot quickly top off a discharged battery pack, it can soak up enough energy in a year to provide 1000 km (621 mi) of free driving energy for customers in southern latitudes, while northern drivers might see a free 500 km (310 mi) of energy from the sun.

Certainly even a small amount of free driving energy helps shift the Tiguan's perception as fuel-thirsty for its size. ■

*Dan Carney*

# 2015 Concept Car Yearbook

# Audi all-electric concept to take on Tesla Model X

**The e-tron quattro concept's sleek styling promises to still look fresh when its production successor arrives in 2018.**

Audi revealed plans for a production battery EV sport crossover in 2018 when it rolled out the e-tron quattro concept at the 2015 Frankfurt Motor Show. The sleek SUV carries a large 95 kW·h lithium-ion battery pack and a trio of electric motors that have a continuous maximum output of 320 kW and a temporary maximum of 370 kW·h and more than 800 N·m (590 lb·ft) of torque.

Though it was termed a concept, "It is not a show car," said Ralf-Gerhard Willner, head of vehicle concepts for Audi. "It is showing the direction for a series production car."

To that end, and to offset the considerable cost of its abundant advanced technology, Audi used as many existing mainstream production components as possible, he said. "It was a goal for the team to use existing components to make it feasible from a finance standpoint."

The concept and eventual production model feature the second iteration of company's MQB platform for longitudinal drivetrains, he said. This "evo" MQB trimmed away 110 kg (243 lb) as it is applied to the latest A4 model, he said. "A 95-kW·h battery is quite heavy, so we had an eye to reduce the weight of the body," said Willner. "To do that, you must know all the components you can size down a little bit."

Audi says the e-tron's driving range is more than 500 km (311 mi) on a charge in highway driving. The vehicle's top speed is limited to 210 km/h (130 mph), and it accelerates to 100 km/h (62 mph) from a stop in just 4.6 s, though testing either of those capabilities will

**Audi will use a vertical array of five OLEDs to power its Matrix laser forward lighting.**

obviously reduce the driving range. The aggressiveness of battery regeneration is driver-selectable though the S and D positions of the shifter and through the Audi drive select system.

A torque control management system sends power to the single front motor and two rear motors to vector thrust for maximum dynamic balance and stability. Additionally, it has all-wheel steering to provide both improved stability and maneuverability as needed.

The hefty battery pack is integrated into the floor beneath the passenger compartment to lower the vehicle's center of gravity and centralize its center of mass. Audi assembles its own battery packs from cells that are manufactured in Europe, but the company declined to identify the battery supplier. Willner did identify the battery chemistry as conventional lithium-ion, though of a next-generation design that is not yet commercially available.

The concept has a dual DC and AC charging system that lets drivers use existing DC chargers or Audi's proposed inductive AC system. It is equipped with the company's self-piloted parking system seen previously that can automatically dock the car atop the inducting charging mat.

In addition to passive measures of sleek, elongated bodywork and a tray enclosing the underbody, the e-tron takes active steps to minimize drag at highway speeds, including an automatically lowering air suspension and movable flaps on the hood, ahead of the wheels, and at the rear to produce a coefficient of drag of just 0.25. That is substantially lower than the class norm of more than 0.30.

Inside, the e-tron concept is a four-seater, with a curved virtual cockpit centered on the driver. The digital instrument cluster is flanked on both sides by organic LED touch screen displays.

The left display controls the Audi's automatic pilot system and its advanced Matrix Laser OLED forward lighting system, while the right handles navigation and infotainment. The car's infotainment system is supplied Internet connectivity via a built-in LTE cellular module.

Audi hasn't mentioned the e-tron's planned production badging, but rumors at the show hinted at the name Q6 to slot it between the existing models. ■

*Dan Carney*

**This artist's rendering of the planned instrument panel depicts the trio of OLED displays spread in front of the driver.**

# Audi shows design, fuel-cell directions

**Audi's Prologue concept sports the company's hexagonal grille that it uses to differentiate cars from the eight-sided grilles on its SUVs.**

At the 2014 Los Angeles Auto Show, Audi exhibited its thinking on the future with a pair of vehicles, one showing its direction for exterior design and the other highlighting the future of electric drivetrains.

The Prologue concept car is a full-size luxury coupe that foretells the company's styling plans for new models above the current A8 sedan, presumably to be called A9.

The concept coupe itself is, at 5100 mm (200 in) in length, slightly shorter than the A8 sedan. But the car's styling is meant to introduce themes that will apply to smaller cars too, and is not intended as a literal representation of one specific future model, according to the car's Chief Designer, Andreas Mindt. "This is a hint for the future," he explained.

For example, the Prologue wears a broad, six-sided grille, which is to be Audi's new standard for cars, while SUVs will have octagonal grilles, he said.

The Prologue's crisp lines are meant to keep Audi's still-familiar styling moving in a contemporary direction, Mindt said. "Audi is a design brand," Mindt noted, and as such must continuously evolve its styling theme.

The Prologue rolls on show-car-typical 22-inch wheels and features Audi's new signature laser headlights. The lasers don't fire directly down the road, because that would obviously be a uselessly narrow beam. Instead, the laser strikes a projection lens that lights the whole street. That beam is supplemented with five smaller lenses fed their light via fiber optics, letting the light adapt to the conditions.

In contrast to some all-aluminum Audis today, the Prologue mixes aluminum with high-strength steel in its construction, perhaps to reinforce strategic areas for crash safety.

At the rear, the LED taillights continue the form language of today's production lights, but underscore the Prologue's width. Fine lines in the light's glass impart a 3D effect when illuminated that creates the illusion of the light moving toward the observer when the brake lights illuminate.

Audi aimed to introduce significant innovations in the Prologue's cabin. Upon entering, the car identifies occupants by their smartphones and adjusts the seating and climate settings accordingly.

The dashboard contains an expanse of three flat-panel touch-screen LED displays that stretch across its entire width. Two are them are in front of the driver, with the left one used to manage lighting and assistance systems, while the right one handles media duties.

The passenger uses the third display panel to individually control entertainment features. The passenger can also set navigation routes, and with the flick of a finger, send them to the driver's display.

The Prologue has a fourth, high-resolution organic LED (OLED) display on the center console, which is used to manage climate control and to input handwriting gestures.

Underneath, the Prologue's significant innovation is the use of four-wheel steering. While this technology has so far found little favor with customers, Audi is confident that its time is coming, according to Ulrich Hackenberg, Volkswagen Group Board Member for Technical Development. "Four-wheel steering will be a typical feature in premium cars like this," he said.

In the Prologue, the rear wheels steer as much as five degrees opposite the fronts for easier parking maneuvers, while they steer in parallel with the fronts at higher speeds for stability's sake.

The car features a 445-kW (597-hp) 4.0-L turbocharged V8 engine that produces 750 N·m (553 lb·ft) driving through an eight-speed Tiptronic automatic transmission to power all four wheels using Audi's quattro technology.

A new wrinkle is the addition of a 48-volt belt-alternator-starter mild hybrid system. It can recover as much as 12 kW of energy under braking and seamlessly restarts the engine when it is off. The company says that the system will soon be introduced in a production Audi model.

Under the more conventional shell of a production A7, Audi also introduced an innovative fuel-cell drivetrain that incorporates a plug-in hybrid electric drivetrain using a lithium-ion battery pack.

Audi has developed fuel-cell technology because of the slow progress in battery development, Hackenberg reported. Fuel cells look like a better solution today, but "the performance of battery cells is increasing," he said. "We have to compare which technology is the winner. We can't answer now."

With currently available technology, the company decided a blend of fuel cells and batteries would be best. The A7 Sportback h-tron quattro concept car's primary drivetrain is a fuel-cell stack powering front and rear electric drive motors that power all four wheels.

Total power is 170 kW, with a combined 540 N·m (398 lb·ft) of torque, which accelerate the 1950-kg (4300-lb) car to 100 km/h (62 mph) in 7.9 s, reaching a top speed of 180 km/h (112 mph). Maximum driving range with full hydrogen tanks is more than 500 km (310 mi), and refilling those tanks takes three minutes, both factors that give fuel-cell EVs advantages over battery EVs.

The fuel cell achieves efficiency of 60%, making it almost twice as efficient as an internal combustion engine. Audi rates the concept car's ability to travel 100 km on a kilogram of hydrogen as the equivalent of 62 mpg.

Audi says the fuel cell is guaranteed to operate at temperatures as low as -28°C (-18°F). Its four carbon-fiber composite outer shell/aluminum inner hydrogen tanks lie beneath the floor of the trunk and together store about 5 kg (11 lb) of hydrogen at 700 bar (10.2 ksi).

**The A7 Sportback h-tron quattro concept has a 170-kW fuel cell underhood that can propel the car 500 km on a tank of hydrogen.**

**Fuel-cell vehicles, like other battery EVs, don't need a tachometer, so the h-tron features a power meter instead.**

With this electric drivetrain in place, failure to recapture energy under braking would be a waste, so the engineering team added an 8.8 kW·h lithium-ion battery pack borrowed from the A3 Sportback e-tron. It not only increases the car's efficiency through the use of recaptured brake energy, it contributes as much as 50 km (31 mi) of reserve driving range in the event there is no hydrogen available.

The plug-in battery pack charges in about four hours using 230-volt service, according to the company.

Hackenberg pronounced the A7 Sportback h-tron quattro's fuel-cell technology ready for production, saying that the sole obstacle to the sale of such cars is the construction of a hydrogen fuel infrastructure to support it. "The first market will be the one with the infrastructure," he said. "Right now, that is California, with Germany not far behind." ■

*Dan Carney*

# GM and LG Corp. partnering for 2017 Bolt EV may create new development model

**LG Electronics Vehicle Components VP Ken Chang (center) speaks with media at the Bolt development announcement Oct. 20. In the foreground is Mark Reuss, GM Executive VP of Global Product Development. At right in background is Denise Gray, the new CEO of LG Chem Power.**

General Motors announced on October 20 that it is partnering with Korean electronics giant LG Corp. in development of the 2017 Chevrolet Bolt EV's entire powertrain and other vehicle subsystems, in a new business relationship that GM engineering executives said could point to a new model for electrified vehicle development.

The GM-LG "joint development agreement" for Bolt involves 11 new and unique components and subsystems, including the battery cells and pack; traction motor (a GM design); power inverter; high-power distribution module; battery heater; accessory power module; power line communications module, infotainment system; and instrument cluster.

Two years ago LG set up a new group, LG Electronics Vehicle Components, that is spearheading the joint effort with GM, according to the group's Vice President, Ken Chang, who spoke with *Automotive Engineering* and other media at GM's Warren, MI, Tech Center. Chang's group includes engineers and technologists from the broader LG family of companies, including battery maker LG Chem; LG Innotek; LG Display; and LG Electronics. He said LG Electronics has invested more than $250 million in a new facility in Incheon, Korea, dedicated to Bolt product development and production.

**Bolt EV's structure is being optimized for light weight to help offset the battery pack's mass. The concept (shown) weighs a claimed 3543 lb.**

The Bolt is scheduled to launch 3Q16 at GM's Orion Township Assembly plant north of Detroit. According to Mark Reuss, Executive VP of Global Product Development, the new electric car is currently in the vehicle integration phase of development, with 100 examples now in road and other testing regimens. Reuss said Bolt will "set new standards for EVs and electrified-vehicle development," with the LG partnership a "disruptive move."

GM has had a growing relationship with LG since 2007, when the Korean company began supplying GM's OnStar module. The companies furthered their bond when LG was selected out of hundreds of battery-cell specialists to supply the lithium-ion pouch-type battery cells for the first-generation Chevrolet Volt, which launched in 2010. Since then LG Chem has delivered more than 23 million cells at less than 2 ppm quality levels, said Pam Fletcher, GM's Executive Chief Engineer for Electric Vehicles.

"This partnership represents a tremendous collaboration, between our engineers and LG's, from top to bottom," Fletcher told *Automotive Engineering*. "Our teams are working shoulder to shoulder."

With LG responsible for so much of the Bolt's powertrain, the relationship appears similar to that of some computer makers with chipmaker Intel. Some industry observers maintain that GM should market Bolt's high level of LG technology content in some way, much like laptop badges that say "Intel Inside." Industry experts note that LG's consumer electronics products are highly respected.

The new GM-LG strategic partnership for Bolt may be the broadest of its kind between a North American automotive OEM and an offshore-based supplier. It is a bold move for both companies, and represents a new way of automotive product development, with potential for speeding time to market and reducing systems cost. The Bolt, with a range greater than 200 mi (322 km) between charges, will be priced from around $35,000 and will compete against Tesla's upcoming (promised for 2017) Model 3 as well as the Nissan Leaf and BMW i3.

According to Reuss, GM's component strategy has three options: build, buy, and partner, depending on which option makes the most economic and strategic sense. ■

*Lindsay Brooke*

# GM's production 2017 Bolt EV to offer +200-mi range at $30K price

**Short front and rear overhangs help optimize cabin space in the Chevrolet Bolt EV concept, the body of which is said to be very close to the production version expected for 2017.**

The race to develop and volume-produce a battery-electric vehicle that is capable of at least 200 mi (322 km) driving range and is priced at $30,000 (after incentives) has intensified, with General Motors confirming on February 12 that it will produce its Chevrolet Bolt EV. Unveiled as a concept at the 2015 North American International Auto Show, the Bolt EV will be built at GM's Orion Assembly plant north of Detroit.

The five-door electric is expected to arrive as a 2017 model, according to suppliers familiar with the project's timeline, although GM has not officially stated the car's start-of-production timing. "We're applying our EV expertise to take electrification further," said GM CEO Mary Barra during the NAIAS. The Bolt will be sold in all 50 U.S. states as well as many global markets.

The car's liquid-cooled lithium-ion battery technology "is in the progression of what we've revealed about the 2016 Volt battery," according to Pete Savagian, GM's General Director of Electrification Systems, who spoke briefly with *Automotive Engineering* during the Bolt's Detroit show debut. While he provided no further details about the Bolt's energy-storage and propulsion systems, it's worth noting that the new Gen-2 Volt uses prismatic-pouch-type cells with a NMC/LMO (Nickel Manganese Cobalt/Lithium Manganese Oxide) chemistry. This enabled a 20% improvement in volumetric energy density. The 2016 Volt's cells are supplied by LG Chem (see http://articles.sae.org/13666/); no word yet on the Bolt's battery-cell supplier.

The concept's exterior form, with its short front and rear overhangs, is "very close" to the production car but the concept interior "is being extensively reworked for production," said a GM engineering source who asked to remain anonymous.
GM engineers are putting a major effort into reducing mass in the Bolt's structure to offset the battery weight; the concept was said to weigh 3543 lb (1607 kg). The production car will use a

**The Chevrolet Bolt EV entering production in 3Q17 at GM's Orion Assembly plant will be very similar to the concept car (shown) unveiled at the 2015 Detroit auto show, according to Mark Reuss, who said the Bolt's range will be above 200 miles per charge.**

**The production Bolt's interior "will be every bit as cool" as the concept vehicle's, shown here, said Reuss.**

mixed-materials strategy consisting mainly of high-strength steel and aluminum alloys, with magnesium and (perhaps) carbon fiber used in specific areas.

If GM actually brings the Bolt to market at $30,000 by 2017, it could be the first EV manufacturer to achieve the feat of combining such a relatively low MSRP with at least 200-mi electric operating range. Tesla CEO Elon Musk has stated that his company's upcoming Model 3, designed for similar range, will enter the market in the 2017 time frame at $35,000 after incentives, Musk said.

By comparison the 2015 Ford Focus EV's base price is also about $30,000 but its 23 kW·h liquid-cooled lithium-ion battery pack delivers a range of 76 mi (122 km) according to the U.S. EPA. Two B-segment EVs priced about $12,000 higher, the BMW i3 and Mercedes B-Class Electric Drive, carry window stickers showing driving ranges of 81 and 85 mi (130 and 137 km), respectively. The BMW's battery pack is rated at 22 kW·h, while the Mercedes' pack is rated at 36 kW·h. The Mercedes is capable of delivering slightly more than 100 mi (161 km) range in ideal thermal conditions.

Bolt's powertrain controller will allow drivers to select operating modes based on preferred driving styles in various operating conditions such as commuting and steady-state cruising. In the concept vehicle, the modes adjust accelerator pedal mapping, vehicle ride height, and suspension tuning. The car also is expected to be capable of dc fast charging.

The concept also featured a dedicated app that enables various functions, including self-parking, to be accessed via smartphone.

To support Bolt production, GM has invested $200 million in the Orion Assembly and nearby Pontiac (MI) stamping plant, with $160 million going to Orion for tooling and equipment and $40 million to Pontiac for dies. The Orion plant is powered by natural gas that is piped in from two nearby landfills. It also uses a 350-kW solar-panel array to generate electricity for use in the plant; excess off-peak energy is sold back to the local power grid. ■

*Lindsay Brooke*

SAE Eye on Engineering Quick Take:

## GM and LG develop new Bolt

General Motors and LG have announced they are co-developing the 2017 Chevrolet Bolt electric car and LG is supplying the entire powertrain, battery, and other electronic subsystems. In this episode of SAE Eye on Engineering, Senior Editor Lindsay Brooke looks at the GM and LG collaboration. SAE Eye on Engineering can be viewed at https://youtu.be/D1-ncd_DcRw. It also airs in audio-only form Monday mornings on WJR 760 AM Detroit's Paul W. Smith Show. Access archived episodes of SAE Eye on Engineering at www.sae.org/magazines/podcasts.

# Challenger GT AWD Concept debuts at 2015 SEMA Show

**The Challenger GT AWD Concept is fitted with Mopar's Scat Pack 3 Performance Kit that includes CNC-ported cylinder heads, hi-flow manifolds, hi-flow catalysts, gasket set, exhaust manifolds, and an exhaust adapter.**

A two-door muscle car prepped for year-round road conditions is the driving influence behind the Dodge Challenger GT AWD Concept.

"The thought for this concept: It's the four-season muscle car. Raise your hand if you'd like to do an all-wheel-drive burnout," Head of Mopar Design Joe Dehner said in an *Automotive Engineering* interview prior to the car's debut at the 2015 Specialty Equipment Market Association (SEMA) show in Las Vegas.

The Challenger GT AWD is one of several vehicles representing the Chrysler, Dodge, Jeep, Ram, and Fiat brands at Mopar's 15,345-$ft^2$ (1425-$m^2$) exhibit. According to Todd Beddick, Mopar's Head of Accessories and Performance Portfolio, "As far as we know, this is the only 8-speed automatic AWD V8-powered Challenger."

The production Challenger R/T's 5.7-L Hemi V8 thumps out 372 hp (278 kW) at 5200 rpm and 400 lb·ft (542 N·m) at 4400 rpm, but the Challenger GT AWD's use of a Mopar Scat Pack 3 Performance Kit elevates the power output by up to 75 hp (56 kW) and up to 44 lb·ft (60 N·m).

"In order to handle all that go and the AWD traction, we added a Mopar prototype lowering kit," said Beddick. The car also has sway bars, performance front and rear strut tower braces, as well as a cold air kit. "And we swapped out the Challenger R/T brakes for the Dodge Charger's police brakes, which gives it much better fade resistance and stopping power," Beddick said.

Compared to the production Challenger, the concept's track was pulled out approximately 2.5 in (63.5 mm) per side. Challenger GT AWD is roughly 2.5 inches lower in the front, and about 2 in (51 mm) lower in the rear. The concept car rides on Pirelli tires and concept Mopar five-spoke forged billet aluminum wheels (20- x 10-in on the front, and 20- x 11-in on the rear).

Header Orange is a body accent strip on the Challenger GT AWD Concept.

According to Dehner, the car features a unique front splitter and a concept rear spoiler. The car's industrial-looking Destroyer Grey exterior paint is complemented by the hood, roof, and decklid being matte black. An asymmetrical stripe sets off the black and grey tones.

"We've done an asymmetrical stripe on a host of different Dodge production and concept cars over the years, but we added something a little different," Dehner said, referencing a 1-in (25-mm) wide Header Orange stripe that seemingly races around the car.

The Challenger GT AWD is a functional car. "That is what's neat about all of our concepts. They all drive," said Dehner.

*Kami Buchholz*

# Divine concept hints at DS brand future

Divine DS shows the future design direction for PSA's DS brand

The evolution of the DS into a brand separate from PSA's Citroën was marked at the 2014 Paris Motor Show by the Divine DS concept, a four-door compact sedan of about the same length as the DS4 at 4.21 m (13.8 ft) long. The Divine DS is designed to show what customers can expect of DS from now on. This will be a combination of sophistication and technology, according to the company. So expect advanced technologies, with elements of French "haute couture."

Head of DS styling Thierry Metroz hopes to have captured elements from the sculptured teardrop designs of French Grand Routiere cars of the 1930s, from designers and coachbuilders such as Figoni and Falaschi, Saoutchik, and Carrosserie Pourtout. "When you look at Divine, the inspiration came from these. With DS, we would like to revive this French spirit," he told *Automotive Engineering*.

Metroz is clear that Divine is just a concept, not a potential future model, but with it, he says that the company is making clear all its future design codes for all future DS models in terms of exterior and interior design as well as color and trim definition. "It's a key concept car in the history of DS," he told *Automotive Engineering*. "[As] with all our concept cars, it's not only a futuristic design vision; we would always like to produce realistic concept cars to show what we could add quickly to production cars. We will see a lot of features or design items from this car on a production car."

Following the DS concepts Numero 9 and Wild Rubis, large concepts designed for the Chinese market, Divine is aimed more at European car buyers who might want a car for the weekend, "It's compact, it's very cute, it's very dynamic," suggested Metroz.

The Divine has been designed without a rear window and in place of a rear-view mirror there is a screen carrying images from a small camera at the rear of the car. "In terms of design, it's a great opportunity because we have no rear window and it's good for our designer to imagine a different design for the rear. The body is also fitted with metal "scales" to give the appearance of a reptile such as a

Double wing symbol will be a signature feature of all future DS models

Divine has no rear window, a camera provides the rear view. "Scales" can be raised and lowered

snake. The scales can be raised or lowered to improve the aerodynamic efficiency of the car. Similarly, the rear lamps are covered by scales, which open up when the engine is started. Metroz suggests that his design team is working on the scales as a production-car feature.

"On Divine, we show our future direction for the front identity of all DS models," said Metroz, "It's a key point when you are working on a premium brand; it's very important to have a strong front and to keep the design for many years and many generations of cars." Referring to the two horizontal wings on either side of the front grille, Metroz said, "All my designers know that you don't touch this—these two wings on the side are our DS signature. All premium brands have a fairly large grille to give status—it's a premium code. But the two wings are unique to DS and it will be a key code for DS on all future models. After that, we are developing a specific design for the grille. To introduce the design we have for the Divine grille on a production car is not so easy." This is because the grille on the concept would not allow enough cooling air to pass through the radiator.

The paired wing signature is already in production in China on the DS5 LS and DS6, having first appeared on the Numero 9 concept in 2012. Metroz said that it will appear very soon in Europe, probably during 2015 and after that it will appear on all new models with the DS emblem, and the Citroen chevrons will disappear. "It's a key point," said Metroz. "You can change the design of the body size or the rear, but it's impossible to change the front every two or three years. If you do that, you are not a premium brand; you're just a brand."

Inside the Divine, DS has invented the concept of "Hypertypage." The idea is to bring the choice of interior trim close to the concept of Paris fashion "haute couture." The dashboard and door scrolls can be changed in 15 minutes to give a choice of three different styles.

These include "Mâle" designed to give a masculine perspective using carbon fiber on black leather with a pearl stitch and red dots.

"'Fatale Punk' is much more rock and roll," explained Metroz, with embossed black leather studded with Swarovski crystals. "The door trim is made with crystal fabric," he added. "It's like a textile, but made with crystals woven in, using microscopic pieces of crystal. It's the first time that the automotive industry has used this type of material. We decided to work together with Swarovski because this company is the world leader for using crystal dedicated to haute couture and technical crystal. Swarovski has a large R&D department with more than 600 people."

The third trim is "Parisienne Chic" using full grain ivory colored leather with pleated silk satin embroidered in the door panels with beads and crystals, produced by the French embroidery designer Maison Lesage.

"So this is our future design philosophy for all DS interiors," said Metroz, "For a concept car we like to deliver strong messages and we exaggerate the design features."

Metroz has other ideas for interiors too, "For example, we are working for a future project to introduce stone on the dashboard. Not small pieces of stone, but big pieces," he said. "We have the technology at the moment to produce very thin, very soft sections. You could, for example, cover all the dashboard with real stone. It feels like stone, has the same visual effect, and it's very light." ■

*John Kendall*

# EDAG CEO discusses the innovative concepts behind the Light Cocoon

**Cocoon prototype in PC illustration.**

The EDAG Light Cocoon was developed with a visionary approach towards a compact, dynamic sports car, with a fully bionically optimized, additively manufactured vehicle structure combined with a weatherproof textile outer skin.

The core idea of the concept is not to regard the body as a closed surface. Instead, an approach was adopted in which material was only actually used in areas where it was necessary for function, safety, and stiffness.

## Simulation on the basis of a series bonnet

To first of all quantify and then verify the lightweight potential of a bionically designed structure, this approach was previously used, for example, in the design of the bonnet of a production vehicle, and then calculated. The point of the calculations was to confirm requirements, e.g. with regard to torsional and flexural stiffness and pedestrian protection.

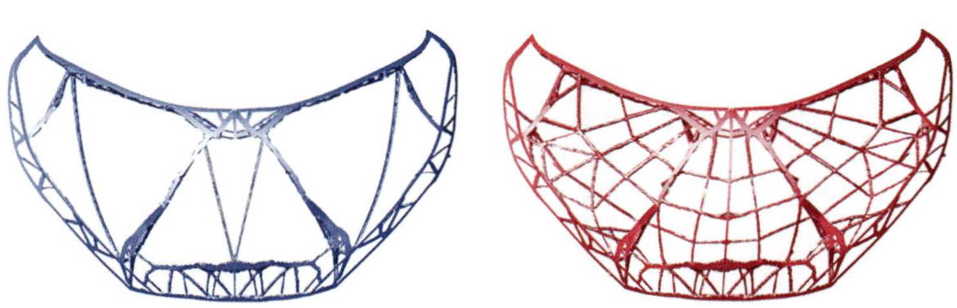

**Bonnet A (left) was not able to withstand the dynamic load cases, especially to meet head-impact specifications, as the hole pattern is too open. Bonnet B (right) shows the geometry that meets head-impact test requirements.**

A very simple geometry for the structure of the bonnet is not able to withstand the dynamic load cases, e.g. to meet head-impact specifications, as the hole pattern is too open. The diagram [in the upper-right-hand corner] shows the geometry that will meet the requirements of the head-impact test.

A topological CAE calculation simulation result confirmed that the basic stability requirements were met, and that a weight saving of about 25% was achieved with this bonnet.

## Fabric outer skin

To protect the vehicle from the effects of the weather despite its open structure, but still meet aerodynamic principles, the leaf of a plant served as the blueprint.

Just like the structure of a leaf, which has a lightweight outer skin stretched over it, a lightweight protective skin covers the Light Cocoon. This textile fabric skin was supplied by the outdoor specialists Jack Wolfskin. The Texapore Softshell O2+ is a triple

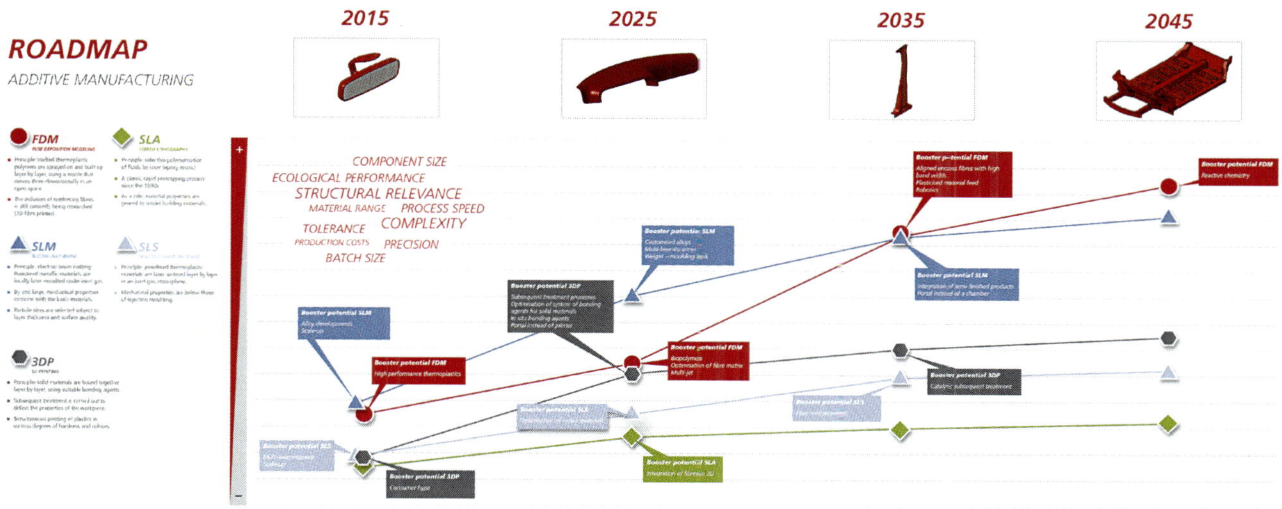


**Additive manufacturing roadmap of Cocoon technology.**

layered polyester jersey fabric that is extremely lightweight, durable, and watertight. Weighing no more than 19 g/m$^2$, the material supports maximum lightweight design requirements with minimum weight.

For the first time it is possible to consider the use of textile materials for the exterior, as these do not have any structural tasks to fulfill. An LED backlight concept creates an among the light, structure, and fabric in an attractive—and above all—innovative vehicle aesthetic.

## Seamless mobility: the Light Cocoon's spoiler

Apart from offering a high degree of lightweight potential, the use of a flexible textile also opens up other chances. Moving components can be covered easily as evidenced by the Light Cocoon's mobile spoiler. The fabric performs every required movement without joins to detract from the vehicle's appearance and is extendable components to generate downforce due to the airflow.

Solutions were created to solve the requirements for the vehicle's spoiler to remain inflexible in its movement, for the cover plate to be bent, and also the length of the spoiler's edge to be adjusted. EDAG experts found a matching PU material able to withstand the tensile stress and deformation. With the help of two infinitely variable servo motors, the spoiler can be electromechanically adjusted to suit the vehicle's speed. Fully extended, the servo is 7 cm (2.8 in) high.

## Personalization via accounts

With the EDAG Light Cocoon, a further dimension is that innovative exterior lighting technology enables the costumer to present his/her vehicle with a variety of lighting setups. With the vehicle, the concept of personalization takes one step further: features are no longer defined in the vehicle; instead, they are linked to a personal account. This makes the account-related activation of selected vehicle functions possible. The driver can, for instance make configurations and settings offline, and then activate them in the car. His/her data are always available without being directly dependent on the car. Also, the cloud-based mapping of some of the functions can transfer computing power away from the car.

## Conclusion

EDAG plans to continue to keep a close watch on the evolution of additive manufacturing. The target: to develop practicable and valid applications for use in component development and production.

A number of selected, structurally relevant components using different materials and additive manufacturing methods were presented at the IAA 2015 in Frankfurt. In addition to this, the company will also continue to press ahead with subjects related to connecting the vehicle to the energy Internet and to the personalized car account.

With the futuristic concept of our Light Cocoon, EDAG wants to make a conscious contribution to stimulating the discussion about the future of lightweight construction and automobile production. As an engineering company, it sees as its task to make sure that it is today already working on technologies of the future and completely new approaches to vehicle development.

When it comes down to it, working on concept car projects like the Light Cocoon enables the product developers to build additional competencies within their company, while at the same time also encourage people to think outside the box and tread new paths to achieve the best possible technical solutions. Companies like EDAG, that develop the cars of tomorrow, must always keep their focus on the day after tomorrow.

*Tobias Meyer, CEO, EDAG Inc., wrote this article for Automotive Engineering.* ■

*Tobias Meyer, CEO EDAG Inc.*

# Infiniti Q60 Concept previews next coupe

**The Infiniti Q60 Concept previews an the all-new engine family, in this case a twin-turbo, direct-injected 3.0-L V6 engine.**

Infiniti revealed its styling targets for its next-generation sports coupe when it pulled the cover off the Q60 Concept in Detroit at the 2015 North American International Auto Show, revealing a machine with exaggerated emphasis on the 21-in wheels and a strong character line running the length of the car.

The car also features the all-new engine family first seen in the Q80 Inspiration concept car at the 2014 Paris Motor Show (http://articles.sae.org/13588/)—a twin-turbo, direct-injected 3.0-L V6 engine whose broad strokes are reminiscent of Nissan's VG 3.0-L twin-turbo seen in the 300ZX sports car.

The concept is constructed with an aluminum unibody structure for weight savings. The company did not explicitly state that the production car will follow the concept's example, but the concept's aluminum body is finished in a glimmering hue named Arctic Aluminum, as if to underscore Infiniti's direction.

The goal, according to Infiniti, is to make the Q60 appear as "a solid piece of liquid aluminum slicing through the air." That character line along the Q60 Concept's sides splits its profile to underscore the car's length.

The concept actually measures 184.6 in (4690 mm) long on a 112.2-inch (2850-mm) wheelbase, and stretches 73.4 in (1865 mm) wide and stands 53.9 in (1370 mm) tall.

The 21-in wheels are specifically selected to draw the eye toward them, with jet turbine-inspired spokes stretching to the rims. "We carefully studied

The Q60 Concept interior takes Infiniti interior design to the next level and appears to be near-production-ready.

The 21-in wheels of the Q60 Concept are specifically selected to draw the eye toward them, with jet turbine-inspired spokes stretching to the rims.

the area of the tire versus the area of the body in the side view, and we concluded to make the wheels more dominant," explained Alfonso Albaisa, Infiniti Executive Design Director. "We'll definitely see more of this in future Infiniti models," he continued.

The Q60's grille also forecasts the brand's direction, he added. "Up front, we show our deep inset, double-arch grille, which floats a bit inside the body, telegraphing its connection to our very powerful engine and serving as more than just simple body decoration," Albaisa declared.

The body contours are meant to evoke chiseled cheekbones and the C-pillar features crescent-cut lines that the company cites as a corporate styling hallmark. "The Infiniti crescent-cut C-pillar is a key Infiniti signature," he said. "Here, the crescent locks into the rearward flow of the roofline, drawing your eye to the graceful, tapering cabin.

Lighting is all-LED, front and rear, with the lenses incorporated as styling elements, while a pair of larger dual exhaust outlets reference the Q50 Eau Rouge concept car.

The powerplant, which drives through a seven-speed automatic gearbox, belongs to a new engine family that will appear in production within two years, according to the company. Infiniti declined to say which car will see first use of the new engine, but did confirm that the production edition of the Q60 concept will arrive in 2016. That production model will closely adhere to the concept's design, with allowances for legal mirrors and practical door handles and wheels. ■

*Dan Carney*

# Q80 Inspiration concept previews future top-range Infiniti, new engine family

**Powering the Infiniti Q80 Inspiration is an all-new, downsized 3.0-L V6 with direct fuel injection and twin turbochargers, coupled with an Adaptive Direct Response Hybrid system.**

Continuing with design themes established with last year's Q30 concept (see http://articles.sae.org/12454/), the Infiniti Q80 Inspiration hybrid concept that debuted at the 2014 Paris Motor Show also showcases what's on the horizon from the luxury brand in the powertrain and connectivity realms.

"Q80 Inspiration is how we see the future in terms of moving people in the higher end of our portfolio," said Francois Bancon, Infiniti Vice President, Product Strategy. "This car is making a statement. Q80 Inspiration is a precursor of a production model in this segment and, as such, integral to our future."

Taking about 15 months to travel from initial designs to the Paris show floor, the four-door full-size fastback is Infiniti's vision of a range-topping luxury sedan. The Q80 Inspiration, which employs both aluminum and carbon composite bodywork, is 5052 mm (198.9 in) long, 2027 mm (79.8 in) wide, and 1350 mm (53.1 in) tall.

Up front, active grille shutters open at lower speeds to cool the engine, and close at highway speeds for better aerodynamic efficiency. Slender full-length hood vents help the engine breathe at higher speeds. Intelligent LED headlights are linked to the car's satellite navigation system and automatically adjust brightness according to the situation.

**Functional C-pillars serve as cool air intakes for the Li-ion battery pack.**

**Portal-style doors create a wide opening to the "1 + 1 + 1 + 1" interior in which each passenger has access to infotainment devices.**

The body design also features "portal-style" doors that create a wide opening to the interior, and functional C-pillars that serve as cool air intakes for the hybrid-boost battery pack.

## Modular engine family, hybrid boost

Powering the Q80 Inspiration is an all-new, downsized 3.0-L V6 with direct fuel injection and twin turbochargers, coupled with an Adaptive Direct Response Hybrid system. The electrified powertrain is expected to produce a combined 550 hp (410 kW)—the V6 providing 450 hp (335 kW) and an electric motor powered by a Li-ion battery pack adding 75 kW of boost when required, including during acceleration to counter turbo lag.

"If the car detects pressure on the gas pedal then it will add to the fun. There isn't just one setting, it'll constantly adjust for the most satisfying drive," said Bancon. "It can also instantly back off the hybrid power boost if the vehicle detects too much speed is being carried into a corner."

The company anticipates fuel efficiency in the range of 5.5 L/100 km (42.8 mpg) and CO2 emissions of about 129 g/km for the hybrid powertrain, which is coupled to a nine-speed automatic gearbox.

The aluminum-based engine will be part of a new engine family expected to be introduced to the Infiniti portfolio within the next two years. The automaker plans to double its powertrain offerings over the next five years.

"The compact V6 engine will replace capacity with a smarter intent—to deliver no-compromise performance and efficiency," said Bancon. "Modern consumers are interested in smart powertrains like this."

Driving the 22-in rear wheels under normal conditions, the vehicle's anticipated 750 N·m (553 lb·ft) of torque can also be split by up to 50% between the front and rear axles. The "smart" drivetrain will switch from rear-wheel to all-wheel drive as conditions dictate, or when manually selected by the driver.

## Philosophy on autonomy—less is more

Infiniti states that the Q80 Inspiration offers a preview of its philosophy on autonomous technology—"that it should enhance the driver and not restrict them." Using cameras, lasers, and sensors, on-demand assistance systems can be employed by the driver when needed through a button on the left-hand side of the steering wheel.

With the car's head-up display (HUD), the driver decides what is projected onto the windshield via an Infiniti InTouch-style touchpad in the center console. The system emits tones suggesting a course of action in heavy traffic or in highway mode, which the driver can heed or ignore.

"Infiniti considers that autonomous driving is an integral part of a premium car of tomorrow. In our case, we have taken the approach that autonomous driving should empower and enhance the driving experience, not do the driving. Our path of development is not taking us into the area of self-driving vehicles," said Bancon.

The "1 + 1 + 1 + 1" interior concept attempts to bring "everything to everyone onboard, when and how they want it." Infiniti designers go about achieving this goal by providing a small, personal touchscreen in the headrests for rear-seat occupants. The front passenger has a HUD for display of external information via a smartphone connected to the vehicle's telematics system.

"There is a tendency with modern design that every time you add interior features and technology you add complexity," said Alfonso Albaisa, Infiniti Executive Design Director. "The idea with Q80 Inspiration was to redefine and simplify how we interact with people in the car. Of course the Q80 Inspiration has the key features which are premium, smart, and engaging, but simplicity was our main objective." ■

*Ryan Gehm*

# Kia reveals electric AWD Trail'ster concept

**The Kia Trail'ster electric AWD concept is more all-terrain savvy than the production Soul with its 5.9 in (150 mm) of ground clearance.**

The Trail'ster concept interprets the Kia Soul as an electric all-wheel-drive compact CUV.

"We keep re-imagining what we could do with the Soul because it's such a popular car. When you have a vehicle like that, you want to make some variations of it. You want to do some different things with it," Tom Kearns, Chief Designer of the Kia Design Center of America, said in an interview with *Automotive Engineering* immediately following the Trail'ster's world debut at the 2015 Chicago Auto Show.

The front-wheel drive Soul arrived in the U.S. market as a 2010 model. A second-generation model followed in the 2014MY, and the all-electric Soul EV was added to the 2015MY product line-up in select U.S. markets.

The Concept Trail'ster sits more than 2.5 in (64 mm) higher than the production Soul. The vehicle's unique fittings include a rollback canvas roof, undercarriage aluminum skid plates, Ksport coil-over shocks, and Pirelli Winter Carving 245/45-19 snow tires.

There are two power sources for the Trail'ster: a 1.6-L turbocharged four-cylinder engine producing 185 hp (138 kW) and 185 lb·ft (251 N·m) that mates to a six-speed automatic transmission and an ac synchronous permanent magnet electric motor producing 27 kW and 100 lb·ft (136 N·m). The 270-volt electric motor, packaged just ahead of the rear axle, and the gasoline engine provide a total output of 220 hp (164 kW) and 285 lb·ft (386 N·m).

The tough-looking appearance of Trail'ster's rear is framed by LED taillamps.

The Kia Trail'ster concept features a rollback canvas roof.

Trail'ster has no mechanical linkages between the engine and the electric motor.

Orth Hedrick, Vice President of Product Planning for Kia Motors America Inc. said in an *Automotive Engineering* Interview that Trail'ster's powertrain configuration is unique. "The gas engine drives the front wheels, and the electric motor drives the rear wheels. That's new for us," Hedrick said.

While the gasoline engine is the primary power source, the electric motor helps provide instant torque during initial startup when the turbo is still spooling up, noted Hedrick, adding, "If the vehicle is on a slippery surface, the electric motor provides additional traction to a second set of wheels."

In a light acceleration mode, the electric motor can drive the Trail'ster for an expected range of 1 to 3 mi (1.6 to 4.8 km).

Trail'ster's electric motor also functions as a generator to recover kinetic energy and recharge the 1.2-kW·h lithium-ion polymer battery pack, stowed under the cargo floor, during vehicle braking or coasting. The vehicle is fitted with an electrically driven A/C compressor, and a hybrid starter/generator for engine stop-start as needed.

Compared to the 2.0-L engine-powered 2015 Soul, the Trail'ster is targeted for a 25-30% fuel economy improvement in city driving and a 5-10% improvement in highway driving.

The concept taps technology from Kia production vehicles such as the 1.6-L turbocharged engine that powers the Forte5 SX. "We have the engine already. We have the electric motor already. We have the battery technology, and we have the platform," said Hedrick, "It's just using the ingredients and coming up with something new and unique." ■

*Kami Buchholz*

LED technology helps create a unique visual identity for the Trail'ster.

# Kia reveals Sportpace concept, three-cylinder

**The Sportpace previews the estate version of Kia's Optima sedan that is more about style than practical load-lugging capability.**

This four-door, four-seater Sportpace previews the estate version of Kia's Optima sedan that will appear in the next 36 months and follows similar styling themes to Jaguar's XF Sportbrake that is more about style than practical load-lugging capability.

Nevertheless it is still a handsome looking car designed under the guidance of Kia's Chief Designer, Gregory Guillaume: "The normal wagon treatment would include a long third window to suggest and show the luggage carrying capacity. But by applying a strong D-pillar treatment and a much more swept back rear hatch, we have given the rear of the car a strong character—muscular and athletic. It has great power and strength within its mass.

"The visual weight of the rear is reduced with careful shaping of screen, door, and rear bumper. The edges cut into the mass to make its raked appearance believable and less wagon-like."

Kia also debuted its all-new 1.0-L T-GDi (turbo gasoline direct injection) Kappa engine. Developed in-house at its Korean Namyang R&D center, the three-cylinder engine develops 118 hp (88 kW) and 126 lb·ft (171 N·m). "The key benefit over the 1.6-L engine," explained head of powertrain, Michael

**New in-house designed and made seven-speed DCT has 221-lb·ft maximum torque capacity.**

**Next-generation wagon will appear within three years as part of the expanded Optima range.**

Winkler, "is that the increased torque is produced at 1500 rpm, whereas the 1.6 only developed 98 lb·ft at the same engine speed.

"The key problem with a four-cylinder engine is that you get an exhaust pulse from a neighboring cylinder when the gas exchange occurs, delivering unwanted gas into the cylinder; with three cylinder engine's firing order this is eliminated."

As is becoming increasingly common with three-cylinder engines, the single-scroll turbocharger—here with a 38-mm (1.5-in) compressor and 34-mm (1.3 in) turbine—is paired with an electric wastegate to improve performance with a highly efficient air-scavenging strategy. As well as scavenging clean air for the engine to re-use for combustion, it is able to open the wastegate at the same time to improve waste airflow.

Direct injection from the intake side at 2900 psi (200 bar) features newly developed, by Kefico, laser drilled injectors with six nozzles in a pyramid pattern delivering an even spread of fuel and air throughout the cylinder while at higher engine loads, with double injections occurring.

The engine also features an integrated inlet manifold and a new dual-thermostat split-cooling system. This allows the engine block and cylinder heads to be cooled independently: the main thermostat controlling the flow of engine coolant to cylinder heads above 190°F (88°C) to reduce knocking, and an engine block thermostat shutting off coolant flow above 221°F (105°C) to reduce mechanical friction and improve efficiency.

Kia also revealed its in-house developed seven-speed, 221-lb·ft (300-N·m) capacity dry dual clutch transmission (DCT) replacing the six-speed automatic and improving economy by up to 7% depending on its application.

The first of its type to be developed in-house, it features a hollow double-gear input shaft that allows the system to quickly engage even gears, while a solid shaft that runs through the middle operates odd gears.

"Downsizing is just one approach to achieve future emissions levels, so I am sure we will have to introduce more technology in the future," said Winkler. "If we look at 95 grams, there is still a long way to go to reduce emissions so we will have to go down the electrification route." ■

*Ian Adcock*

**Kia is the latest manufacturer to join the trend of going to downsized, three-cylinder direct injection and turbocharged gasoline engines.**

# Lexus unveils A-segment crossover concept

**Lexus LF-SA Concept is a 2+2 A-segment crossover concept.**

Crossover concepts were not in short supply at the 2015 Geneva Motor Show, particularly from Japanese manufacturers, and angular design seems to be the trend this year. Lexus' LF-SA concept was undoubtedly the smallest and the product of Lexus ED2 design studio based in Nice, France.

This is the first complete Lexus to have been designed at ED2. The concept was designed to help celebrate 25 years since Lexus launched its first car, the LS400. It is a design study exploring new styling directions for Lexus, commented a Lexus spokesman.

The LF-SA stands for Lexus Future Small Adventurer and the exterior, finished in Stellar silver, is said to be a reference to space exploration. Lexus described the concept as an ultra compact, sub-B-segment urban 2+2, measuring 3450 mm (135.8 in) long, 1700 mm (66.9 in) wide, and 1430 mm (56.3 in) tall. This makes it 5 mm (0.2 in) shorter, 85 mm (3.3 in) wider, and 30 mm (1.2 in) lower than the Toyota Aygo A-segment city car.

**The concept is the first complete car designed by Lexus ED2 design studio in Nice, France.**

It is recognizable as a Lexus from the front by the adoption of the latest shape Lexus ripple finish grille, although in the case of the LF-SA the ripples are radial, extending from the central Lexus badge on the grille. The coupe body and three door format gives it the appearance of a two-seat design, although two occasional seats are provided in the rear.

The design incorporates convex and concave surfaces with the pronounced crease picked up from the lower section of the A-pillar and flowing into the edge of the hood. From the rear a deeper crease picks up the beltline, bringing it back forward with the crease flattening under the lower edge at the front of the doors.

The result is a subtle blend of what would normally be sharply contrasting shapes. The shape of the front grille is echoed in the rear panel of the car. Features such as the L-shaped rear lamp clusters and Lexus arrowhead motif reflected in the shape of the daytime running lights are design features that can be seen in current Lexus production models.

Lexus insisted that the LF-SA is not an autonomous vehicle design study, suggesting that it is a driver-focused vehicle where drivers can enjoy the luxury of driving themselves, while still being able to experience the latest in infotainment, connectivity, and safety.

The driver-focused theme is emphasized inside the car by the adoption of unique upholstery colors for the driver and front passenger seats. This is further reinforced by giving the driver a seat with a fixed position but adjustable pedals and steering wheel. Rear seat passenger access is from the passenger side where the front seat will slide to permit entry. Instrumentation includes a wide-angle head-up display and a hologram-inspired digital display for the infotainment system.

LF-SA is intended purely as a design study and not a concept for a future production model. No indication of a possible powertrain was given. ■

*John Kendall*

# Supercar concept previews Lamborghini plug-in possibilities

**The Lamborghini Asterion LPI 910-4 is built using a carbon-fiber monocoque**

The Lamborghini Asterion LPI 910-4, revealed at the 2014 Paris Motor Show, demonstrated how the Italian company could produce a plug-in hybrid supercar concept using technology available today. The Asterion combines the naturally aspirated V10 engine of the Huracan with three electric motors to provide a powerful gasoline/electric powertrain with low carbon dioxide emissions.

Lamborghini explains that the Asterion LPI 910-4 name is derived from the following: "LP" for "longitudinale posterior" indicating which way the engine is mounted (longitudinally) and its position on the car (behind the cockpit), I indicates the Italian for hybrid "ibrido," 910 represents the system power by adding the power output in horsepower of the V10 engine and the electric motors together, and 4 representing the four-wheel-drive capability of the drive system.

The V10 engine has a displacement of 5204 $cm^3$ and produces maximum power of 610 hp (455 kW) at 8250 rpm and peak torque of 560 N·m (413 lb·ft) at 6500 rpm. The engine is positioned longitudinally in a mid-engine location behind the cockpit, driving the rear wheels through a seven-speed dual-clutch transmission in a transaxle arrangement.

The first electric machine is located between the V10 engine and the dual-clutch transmission and functions as an integrated starter/generator (ISG). Two further electric motors are located in the front axle and fed by the ISG system, which incorporates

**Asterion hybrid power involves a Huracan 5.2-L V10 combined with an ISG and two electric motors driving the front wheels, giving three drive modes.**

torque vectoring to the front wheels. The combined power output of the three electric motors is a claimed 220 kW. The lithium battery pack is fitted longitudinally in the "transmission" tunnel, with the aim of improving vehicle balance and protecting the batteries in the event of a lateral impact. Lamborghini claims that the weight of the hybrid technology is 250 kg (550 lb).

This drive system has three modes. In hybrid mode, the V10 engine and all three drive motors are engaged. These function without reliance on the state-of-charge of the batteries to provide permanent four-wheel drive. In electric-drive mode, only the front wheels are driven by the two electric motors, and the batteries provide a claimed range of 50 km (31 mi). In thermal engine power mode, only the V10 engine is used to drive the rear wheels.

Claimed performance includes acceleration from 0-100 km/h (0-62 mph) in 3.0 s, a top speed of 320 km/h (200 mph), and a top speed of 125 km/h (78 mph) under electric power alone. Tested on the European New European Driving Cycle (NEDC), the car is said to return 4.2 L/100 km, or about 56 mpg U.S., on the combined cycle test, with carbon dioxide emissions of 98 g/km.

The Asterion is based on a carbon-fiber monocoque to assist in reducing weight. The front end of the car is constructed as a single component. Materials used for the headlamps include forged carbon fiber and titanium. The Pirelli tires are fitted to 20- and 21-in rims that include carbon fiber components.

Inside the car, aluminum, titanium, and forged carbon fiber are among the materials used. Titanium is used on the steering wheel, whose design is drawn from the Lamborghini Miura. The three drive-mode buttons are incorporated in the steering wheel. Climate control, GPS, and the infotainment system are among the functions controlled from a portable tablet computer. ■

*John Kendall*

# China luxury market requirements drive new Lincoln Continental design, engineering

Lincoln Continental Concept accurately points to production car launching in 2016.

"This concept is a strong indication of what the production model will look like," said Ford Motor Co. CEO Mark Fields, at the New York media introduction of the Lincoln Continental concept on March 29. "It's the 'next chapter' of Lincoln."

In development for two years, the all-wheel-drive Continental is scheduled to enter production in 2016 on Ford's CD4 flexible architecture that underpins Taurus and is engineered to support AWD and FWD drivelines, according to a key supplier involved with the program. The Continental debuts a bold new design language for Ford's premium brand. Gone is Lincoln's signature "waterfall" split-grill front end. Inside, the Continental's enormous and sumptuous rear-passenger cabin area was defined by requirements in the booming China luxury market, where discriminating owners are driven in chauffeur fashion rather than driving themselves.

The car's overall form and details are so bold, in fact, that the concept seemed to catch off-guard most of the reporters assembled for the car's unveiling in Manhattan. If placed in any Lincoln showroom today, the handsome and exquisitely trimmed

**Front-seat cabin area of the Continental Concept is close to production-spec. The new Lincoln flagship is expected as a 2017 model.**

Continental would immediately obsolete the rest of the current portfolio. It's a significant directional statement by Lincoln Design Chief David Woodhouse.

"We have to move the brand forward and it can't be with baby steps," Fields asserted. "Lincoln is truly core to our growth plans." He noted that sales of luxury vehicles are responsible for about 33% of the industry's profits.

Ford's $5B investment in Lincoln's rejuvenation is aimed at tripling volume from the 100,000 units produced in 2014 within the next few years. It is also providing dedicated design and engineering resources, as well as technology features, to support Lincoln's ambitions to become a global premium player that is decoupled, in the public's eye, from the Blue Oval.

"We're very focused on how customers experience the vehicle—and how we deliver that experience will be unique to Lincoln," said Kumar Galhotra, the Lincoln President and formerly Ford's Vice President of Engineering and veteran Asia-Pacific product planning chief.

As examples, he listed the Continental's new 30-way-adjustable seats; E-Latch door handles that open the doors by touch rather than by pulling and are integrated into the car's upper beltline; Lincoln-exclusive 3.0-L EcoBoost V6; active noise cancellation; and its audiophile-quality Revel sound system (co-developed by Harman International in a 10-year-exclusive deal with Lincoln) as "technologies that will define the Lincoln Motor Car Co."

The new seats use a new synthetic cushion material that contributes to structural strength. According to Scott Tobin, Director of Product Development, the seats were designed in house and have 50 patents either applied for by Lincoln or granted. They feature independently controlled left and right thigh supports to increase driver and passenger leg comfort particularly during long trips.

The car also will feature a tailored version of Ford's 360° camera-and-radar-based parking-assist system. Another technology on the concept that is expected to make it to production is SPD SmartGlass, a product of Research Frontiers, Inc. SmartGlass is used in the concept's tinting sunroof and allows

**The new Continental's poshly trimmed rear passenger cabin is aimed at the "driven customer" in China. Leg room is extraordinary and the car's rear doors are lengthened for optimum ingress/egress.**

**LED matrix headlamps feature laser-assist high beams. The Lincoln 'cross' badge serves as a visual centerpiece of each lamp.**

**Lincoln Director of Product Development Scott Tobin said the new 30-way-adjustable seats, developed in house, carry 50 pending or granted patents.**

**Lincoln Motor Car Co. President Kumar Galhotra knows the China market extensively.**

One of the Continental Concept's Revel audio speakers developed by Harman International, mounted in the front passenger door.

Novel E-Latch door handles operate by touch rather than pulling force.

passengers to control heat from ambient sunlight. Ford engineers say the glass can cool the cabin by as much as 18°F (10°C) while blocking 99% of UV rays.

But it's that cavernous and posh rear-seat cabin that will most distinguish the new Continental from Lincoln's current "large" sedan, the D3-based MKS that is scheduled for retirement within a year. Back-seat passengers are enveloped by high-fashion materials that tastefully juxtapose aromatic, buttery-texture animal hide (and convincing faux hides) and real wool carpet with polished metal accents.

SAE Eye on Engineering Quick Take:

## Cadillac vs. Lincoln

Watch the video at https://youtu.be/KzKcvY8wiKg

Cadillac vs. Lincoln is one of the world's great automotive rivalries. In this episode of SAE Eye on Engineering, Senior Editor Lindsay Brooke looks at the new Cadillac CT6 and Lincoln Continental. SAE Eye on Engineering can be viewed at https://youtu.be/KzKcvY8wiKg. It also airs in audio-only form Monday mornings on WJR 760 AM Detroit's Paul W. Smith Show.

Access archived episodes of SAE Eye on Engineering at www.sae.org/magazines/podcasts.

"There is a balance we had to strike between creating the 'driven' car for China and more of a driver's car for the North American market—and also for China," Galhotra told *Automotive Engineering*. "It's important for the Chinese customers who are being driven to have lots of space and all kinds of amenities—music, multi-adjustable seats—that they can control from the rear-seat area. But many of these luxury customers also want to go out and drive their cars themselves on weekends.

"It's a young demographic, in their 30s and 40s, and many of them want an engaging vehicle as well as a comfortable, quiet, and effortless vehicle," he noted. "That's what we have to deliver."

According to Galhotra, the Continental's main competitors include the BMW 5 Series, Audi A6, and Lexus GS—midrange sport-luxury rather than flagship products. The Lincoln appears to be sized between the global D- and E-segments (Ford has not yet released dimensions or other specifications).

He said a long-wheelbase version has been under investigation. ■

*Lindsay Brooke*

# Mercedes' concept stresses connectivity, record-breaking aerodynamics

**A gasoline/electric plug-in hybrid drive with a total output of 205 kW (275 hp) provides the Concept IAA with an electronically limited top speed of 250 km/h (155 mph). In aerodynamic mode, the vehicle manages an all-electric range of 66 km (41 mi) and emits 28 g/km of CO2.**

Among its five world premieres at the IAA (Frankfurt Motor Show) 2015, Mercedes-Benz surprised show-goers with the Concept Intelligent Aerodynamic Automobile (Concept IAA), which the German automaker also calls the "Digital Transformer." The concept car is designed to show how far advanced Mercedes-Benz already is in the "digitalization" of automotive development and production, according to Dr. Dieter Zetsche, Chairman of the Board of Management of Daimler AG and Head of Mercedes-Benz Cars.

"The Concept IAA shows that the real and the virtual world are merging more and more at Mercedes-Benz," said Zetsche, speaking at the concept's Frankfurt Motor Show reveal. "Never before have we developed a vehicle concept as quickly as our Concept IAA. What previously took up to one and a half years, we managed in less than 10 months thanks to digitalization."

The four-door "coupe" is 5040 mm (198.4 in) long—5430 mm (213.8 in) long when in "aerodynamic mode" (see below)—1995 mm (78.5 in) wide, and 1305 mm (51.4 in) high. Its wheelbase measures 2975 mm (117.1 in), and the front/rear track widths are 1710 and 1770 mm (67.3 and 69.7 in), respectively.

A gasoline/electric plug-in hybrid drive with a total output of 205 kW (275 hp) provides the concept car with an electronically limited top speed of 250 km/h (155 mph). In aero mode, the vehicle manages an all-electric range of 66 km (41 mi) and emits 28 g/km of CO2. In shorter "design mode," the range is 62 km (38 mi) and CO2 emissions are 31 g/km.

## Active aerodynamics

The Concept IAA is a world record-breaker for aerodynamics, according to Mercedes, with a Cd figure of 0.19. At around 80 km/h (50 mph), the vehicle automatically switches from design mode to aerodynamic mode, changing its form with a number of active aerodynamic features.

**The interior features a "touch-based" operating philosophy. The center console features a trim element of curved glass in which a touch display is integrated.**

At the rear end, eight segments made of carbon-fiber-reinforced plastic (CFRP) deploy to extend its length by up to 390 mm (15.4 in), "substantially" reducing the after-flow zone behind the vehicle and thus drag. Flaps in the front bumper extend outward by 25 mm (1.0 in) and rearward by 200 mm (7.9 in), improving airflow around the front end and the front wheel arches. The fin in the front bumper retracts by 60 mm (2.4 in) to improve flow along the underbody. In addition, active wheel rims change their cupping from 50 mm (2.0 in) to zero—from five-spoke to flat-disc wheels.

**Active wheel rims change their cupping from 50 mm (2.0 in) to zero—from five-spoke to flat-disc wheels.**

As a result of this transformation, the Cd value improves from 0.25 to 0.19. The frontal area totals 2.16 m$^2$ (23.3 ft$^2$). Because the downward slope of the concept car's roofline begins further toward the front, designers incorporated two "rises" over the rear seats to offer the rear passengers sufficient headroom.

Other aerodynamic optimizations include windows that fit flush on the outside, a blue LED touchpad instead of door handles, and lowering of the chassis, resulting in a ground clearance of 100 mm (3.9 in). The underbody paneling partially covers the center tunnel, with perforations to allow cooling of the exhaust system, and extensive cladding on the rear axle. Cameras in the side air outlets behind the front wheel arches project images onto the split-screen rearview mirror in the interior, making exterior mirrors unnecessary.

As on the new C-Class, an adjustable radiator grille shutter (air panel) helps to reduce drag. When only low cooling requirements apply, the concentric louvers in the radiator grille are closed to prevent air from entering into the engine compartment. Basic ventilation then occurs primarily via the Mercedes star and the cooling air opening in the bumper.

The aerodynamic features were developed with the aid of numerical flow simulation. The automaker's aerodynamics experts used around one million CPU hours to simulate the airflow, working through

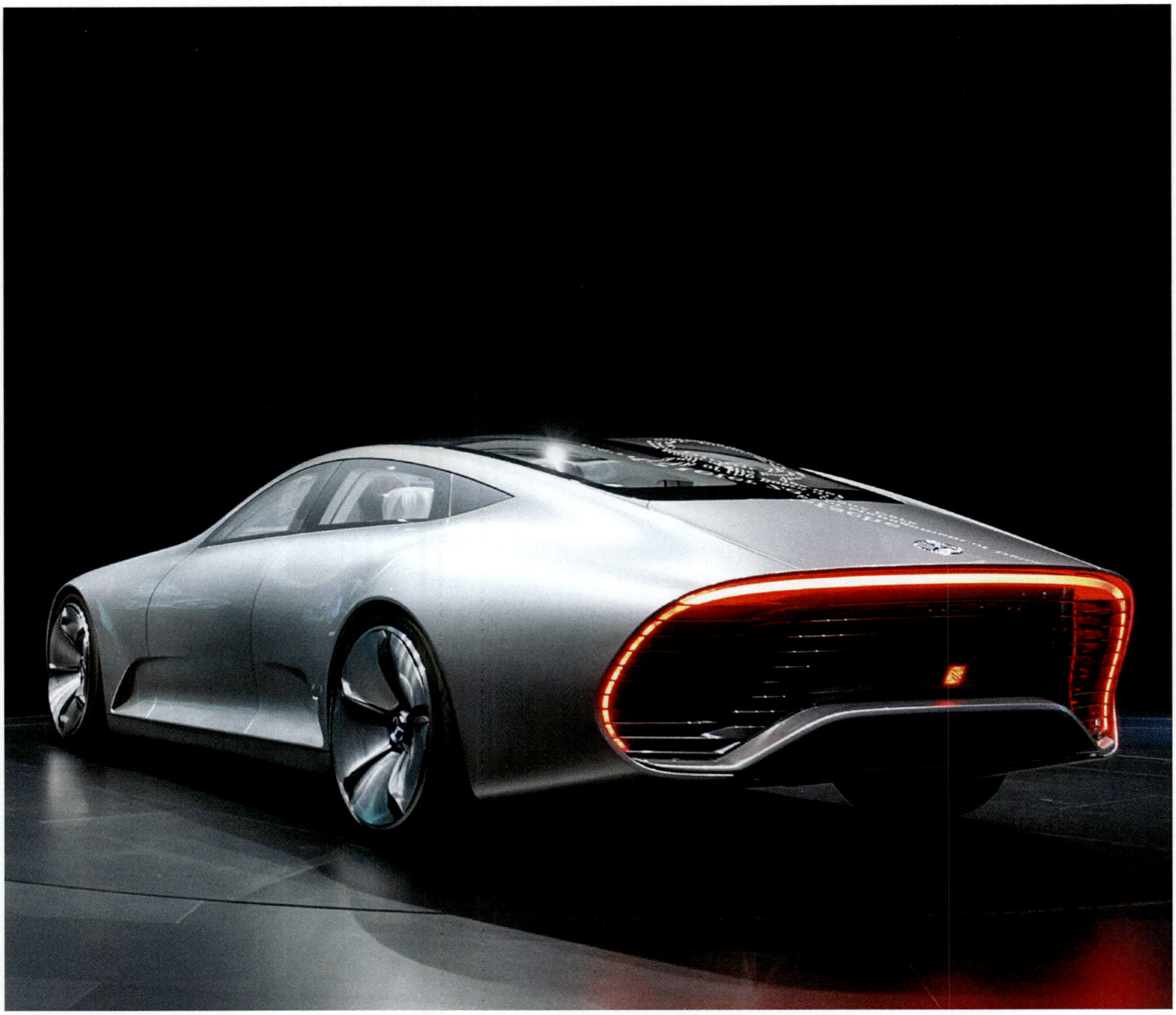

**Aerodynamic optimizations include windows that fit flush on the outside, a blue LED touchpad instead of door handles, and lowering of the chassis, resulting in a ground clearance of 100 mm (3.9 in).**

around 300 variants. The work involved is roughly equivalent to that required to develop a production model, according to Mercedes. Fine-tuning then took place in the wind tunnel in Sindelfingen.

"The Concept IAA applies intelligent innovations to resolve the conflicting aims of functionality and aesthetics and shows that we still have plenty of ideas on how to achieve further improvements in efficiency," said Prof. Dr Thomas Weber, Member of the Daimler Board of Management responsible for Group Research and Head of Mercedes-Benz Cars Development.

## Digitalization and connectivity

In addition to impressive aerodynamics, the Concept IAA reflects the "fundamental technological changes" taking place in the automotive industry, for which the primary driving force is "digitalization, also known as Industry 4.0." According to Zetsche, a continuous digital process chain from research and development through production to marketing and sales, logistics, and service is already becoming a reality at Mercedes-Benz.

The design and aerodynamic shaping of the concept was aided by the digital networking of different specialist departments. Designers employed the latest algorithmic design methods to handle the complex geometries, which were turned into reality via production technologies such as rapid prototyping.

The concept car also is equipped with a large number of sensors and modules that enables autonomous driving and car-to-x communication. Inside, the Concept IAA continues the design theme of the S-Class and S-Class Coupe. New touch-based

**The Concept IAA boasts a Cd figure of 0.19. Active aerodynamic features include eight CFRP (carbon fiber reinforced plastic) segments at the rear that deploy to extend the car's length by up to 390 mm (15.4 in), reducing drag.**

functions hint at what the interior of a business saloon might look like in the near future, Mercedes says.

The automaker worked with Silicon Valley-based Nvidia and The Foundry, a creative-software developer for visual effects and games, on the in-vehicle digital displays.

The instrument cluster consists of two 12.3-in screens positioned side by side. The displays are flush-mounted behind glass. The glass elements are all bonded, which involves filling the minimal gap between front glass and film with a transparent adhesive in the production process. This provides for substantially improved optical quality and robust screens, according to Mercedes.

The center console features a trim element of curved glass in which a touch display is integrated. This is where air-conditioning and seat-adjustment functions are operated, and where entry buttons for the different operating menus are located. The exterior aerodynamic elements can be controlled in a new menu.

The two-spoke steering wheel also incorporates touch-based operating functions: OFN (Optical Finger Navigation) buttons, embedded in the clusters in a similar hovering manner as the touchpad in the center console, allow the driver to scroll through the instrument cluster menus. The OFN button on the left controls the left-hand display, the button on the right controls the right-hand display.

Trim elements are all-aluminum, and the entire floor of the show car is covered in leather. The eight air vents have been produced by Swarovski, accented by blue and red ambient lighting. ■

*Ryan Gehm*

# Mercedes-Benz debuts 'car of the future' at CES

**"This is the Mercedes-Benz vision of the next automobile revolution," said Dr. Dieter Zetsche, Chairman of the Board of Management of Daimler AG and Head of Mercedes-Benz Cars. "Cars will be re-thought in terms of shape and design, in terms of outside interaction, and in terms of inside interaction."**

Designed around the future luxuries of space, time, and privacy, the Mercedes-Benz F 015 "Luxury in Motion" concept car had its world premiere Monday night at the 2015 International CES.

"The idea is about designing the entire car around the advantages of autonomous driving, while making use of the added room you have in an electric car without the engine and transmission," said Dr. Dieter Zetsche, Chairman of the Board of Management of Daimler AG and Head of Mercedes-Benz Cars.

Providing the maximum possible space for the passengers was achieved with a large wheelbase of 3610 mm (142 in) and short overhangs. Length, width, and height are deemed "unusual proportions" by Mercedes-Benz with 5220, 2018, and 1524 mm (205.5, 79.4, and 60.0 in).

"We were able to create the monolithic appearance of the exterior—as if hewn from a single piece—in particular through doing without conventional windshield and window design," said Gorden Wagener, Vice President Design, Daimler AG. "The flush-fitting side windows have an almost mirror-like appearance thanks to their extremely finely patterned surface. As they are close in color to the cool, technical appearance of the alubeam paint finish of the vehicle body, the transition from the windows to the bodywork is almost imperceptible. In addition, there are neither exterior mirrors nor visible A-, B-, or C-pillars to interfere with the overall impression of a sensual, object-like sculpture."

The interior features a variable seating system, with four rotating lounge chairs that allow a face-to-face seat configuration. Vehicle windows can be used as screens, with six display screens integrated into the instrument panel and the rear and side panels. Passengers can interact with the vehicle through gestures, eye-tracking, or by touching the screens. A free-floating control unit enables any of the four passengers to take control of the vehicle.

To ease ingress and egress, the electrically powered seats swing outwards by 30° as soon as the doors are opened. With rear-hinged saloon doors, the front and rear doors can be opened and closed independently of one another.

The F 015 has large communication displays at the front and rear with LED panels. By changing the color of the lights, the vehicle indicates what driving mode it is currently in: blue for autonomous and white for manual.

"In the interior we focused on comfort, an extraordinary sense of space, and the luxurious lounge character," Wagener said. "The displays allowing nonstop, all-round information and interaction link the analog and digital worlds together."

The high-resolution screens in the side windows can be used to display personal information for each passenger, including Web browsing, contacts, email, and video conferencing.

The B-pillar has been eliminated thanks to a specially structured body. A sturdy interconnected system with mechanical locking elements ensures safety standards are met. The resulting composite load path allows a high amount of energy to be absorbed in a frontal or side impact with minimal intrusion into the passenger compartment.

New materials and structures were used to develop the Smart Body Structure of the F 015. With the strategic use of carbon-fiber-reinforced plastic (CFRP), aluminum, and high-strength steels, the bodyshell was able to achieve a 40% weight reduction, compared with today's production vehicles.

**The interior features six display screens integrated into the instrument panel and the rear and side panels.**

Steel side members have been replaced in favor of plate-shaped multi-chamber sections made from CFRP composite. CFRP is also used for the pressurized hydrogen tank system's cylinders and mounting system.

Large LED light modules are positioned at the front and rear, which provide a range of different lighting functions. Communicating with the outside world by means of the LED fields, the lights show if the vehicle is being driven autonomously (blue) or is controlled manually (white). Road safety is also increased via a unique pedestrian communication system.

"Should a pedestrian wish to cross the road, for instance, the F 015 halts and scans its surroundings to check whether it is safe to do so," said Prof. Dr. Herbert Kohler, Head of Corporate Research & Sustainability and Chief Environmental Officer for Daimler AG. "If it is, the car uses a high-precision laser system to project a virtual crosswalk onto the road surface and additionally lets the pedestrian know it is clear to cross with an audible 'Please go ahead' prompt."

The F 015 was designed to allow the integration of an electric drive system with a fuel cell. The electric hybrid system has a total range of 1100 km (684 mi), including around 200 km (124 mi) of battery-powered driving and around 900 km (559 mi) on the electricity from the fuel cell.

The fuel cell stack delivers the traction current for two electric motors, each with an output of 100 kW. The electrical drive system generates a total peak power output of 200 kW (272 hp) and maximum torque of 200 N·m (148 lb·ft) per motor is available instantly. The F 015 is said to go from 0 to 100 km/h (0 to 62 mph) in 6.7 s with a governed top speed of 200 km/h (124 mph). ■

*Matthew Monaghan*

# Mitsubishi PHEV concept destined for production?

**Mitsubishi's Concept XR-PHEV II could point the way to a future compact plug-in hybrid crossover.**

Mitsubishi hints that the Concept XR-PHEV II unveiled at the 2015 Geneva Motor Show could be the basis of a second plug-in hybrid electric vehicle (PHEV) model, following the Outlander PHEV which reached the market in 2012.

The concept is described as a small SUV, but it differs significantly in the PHEV drivetrain department compared with the Outlander. While the serial hybrid Outlander PHEV drivetrain can deliver electric drive to all four wheels, the Concept XR-PHEV II has been designed as a front-wheel-drive only crossover. Internal combustion power is provided by a Mitsubishi MIVEC engine, featuring variable valve timing and lift. Drive torque supplied by an electric motor is also directed through the front wheels.

Mitsubishi says that the drivetrain is designed to propel the vehicle using electric drive for most of the time. The system would switch automatically between series and parallel hybrid operation. Either the engine would drive the generator to produce electric power when needed, or the battery provides power directly to the drive motor. In parallel hybrid operation, the engine would provide drive power directly to supplement the operation of the electric drive system.

Few further details were provided about the hybrid system, but Mitsubishi claims that it is a lightweight and compact design. The drive unit is said to include a high-efficiency inverter and variable voltage system to give high power and torque.

The drive system provides Battery Save and Battery Charge modes. Battery Save uses the internal combustion engine to drive the generator, thereby providing electric power directly to the drive motor

**Concept XR-PHEV II would use a front-wheel-drive hybrid powertrain.**

and retaining battery charge for use in zero emission zones. In Battery Charge mode, the generator output is used solely for charging the 12-kW·h battery pack. Mitsubishi claims an electric drive motor output of 120 kW and carbon dioxide emissions that could be below 40 g/km, although no test cycle is referenced.

Concept XR-PHEV II measures 4490 mm (176.8 in) long, 1890 mm (74.4 in) wide, and 1620 mm (63.8 in) tall, on a 2670-mm (105.1-in) wheelbase. It has recognizable crossover design cues including the raised ground clearance, short overhangs, large alloy wheels, and wheel arch cladding. Prominent skid plates are fitted front and rear. The design contrasts curves with sharply angular sections. The rising beltline is echoed by the deep crease below it, partially concealing the door handles. The crease is also continued in the hood section, tapering down to the angled front of the vehicle. Narrow-section door "mirrors" suggest the use of rear view cameras.

The interior is laid out with a four-seat configuration, featuring a broad center console extending from the front to the rear of the cabin. Seats and center console were trimmed mostly in orange upholstery. The full-length center console is echoed in a similar panel in the glass roof.

The instrument panel is supported by an aluminum frame which is also repeated in the glass roof sections, giving the interior the appearance of an aircraft cockpit with some switchgear in an overhead console. A monitor screen is located here too, with another rising from the center of the dashboard. Most switchgear is situated in the center console and on the steering wheel. ■

*John Kendall*

# Nissan shows Gripz crossover concept

**The Gripz rides on airy 22-inch wheels developed especially for it by Bridgestone to evoke bicycle wheels.**

After a run of success with unorthodox SUV styling, especially in the Juke subcompact crossover SUV, Nissan has apparently decided to mine this vein as the company's new signature.

The result, shown at the 2015 Frankfurt Motor Show, is the Gripz Concept, a sleek crossover SUV that seeks to chisel another niche into the Nissan's already-packed line of SUVs. "We want to redefine the crossover segment," explained Giovanny Arroba, Creative Design Manager for Nissan Design Europe, the London studio responsible for the concept.

Citing the Juke as a prior success in this effort, he said, "This is a very important segment for us, and we wanted to see how far we could push it."

Far enough to ruffle enthusiasts' feathers, who fear that the Gripz could be approved for production at a time when the 370Z may be dropped from the line, prompting them to view the Gripz as the Z-car's replacement. Arroba insists that the Gripz stands on its own and is unrelated to other product decisions,

**The Gripz' color scheme was inspired by the Nissan Safari Rally Z of the 1970s with its matte black hood.**

but the company made enough attempts to link the Gripz to the Z that suspicions remain that Nissan views it as a next-generation Z.

Aside from the "z" at the end of its name, the Gripz Concept also wears a matte black hood with orange-red paint as a tribute to the Safari Rally Z, an off-road racing version of the early 240Z. Nissan won the Safari Rally with the car.

Aside from the nod to the off-road sports cars of 40 years ago, the main inspiration for the Gripz Concept was road bicycles, reported Arroba. The car features a carbon frame, with body panels mounted over it. Exposed carbon elements such as the boomerang-shaped door structure underscore its construction.

"The interior guys brought in the road-bike inspiration, and it spread to the exterior because it struck a chord with us," said Arroba. The idea was to leave frame materials and vehicle componentry exposed as they are on bicycles, he said. "We're not padding and insulating and hiding that," he said.

Beyond the openings for the scissor front and rear-hinged rear doors, sharp angles abound in the Gripz Concept, from its V-shaped grille to its indented triangular sills. Even the exhaust pipe tips are trapezoidal.

It does have exhaust because there is an internal combustion under the hood, though the concept's powertrain is a series hybrid, so the gasoline engine's power spins a generator that powers an electric motor borrowed from the Leaf EV.

Nissan provided no additional technical details, just saying that the electric drive system lends the concept a smooth, refined, and exhilarating driving experience with outstanding fuel efficiency. ■

*Dan Carney*

# Peugeot Fractal demonstrates a new take on sound design

**Peugeot Fractal concept can be accessed using a smart watch, which triggers a "sound signature" when the vehicle is unlocked.**

Peugeot used the 2010 SR1 concept to introduce its i-Cockpit touchscreen-based system, designed to control heating and ventilation, navigation, audio, connectivity, and the trip computer. The first-gen system entered production in the 2012 Peugeot 208. The company took the i-Cockpit concept one stage further in the Fractal electric urban coupe concept at the 2015 IAA Frankfurt Show, to include sound.

The concept plays a "sound signature", created by DJ and sound designer Amon Tobin, which is triggered when the driver opens the car using the smart watch remote locking system. "Behind this concept of an electric car—that is not really new, there was this question that electric cars emit no sound", explains Matthias Hossann, Head of Concept Cars and Advanced Design at Peugeot.

"In France, sometimes if you say you are driving an electric car, people say they are very sorry for you because they think it is not as exciting to drive an electric car. We thought about that and agree that you lack some pleasure with an electric car because it emits no sound. I have had this experience because even when you start an electric car, you don't know that it has started."

The Peugeot design team's thinking was that sound is part of the conventional driving experience, giving the driver an indication of speed, for instance. "We think that by developing some specific sounds, we can generate some specific emotions, some specific driving pleasure and also a more precise driving experience", explains Hossann.

As a result, Peugeot has developed some more specialized sounds for the Fractal. "The best example is the GPS," continued Hossann. Most of us have had the experience on a busy road of missing a turn even though the navigation system has given the instruction. The Fractal designers have used directional sound to help. "We think that by developing some sounds that come from one side, for example the right, with a command to turn right in 300 m, followed by some other sound from the right, will give a better driving experience."

**More than 80% of the Fractal interior is manufactured using 3D printing.**

The Peugeot design team has worked with Tobin since the beginning of the project to design specific sounds. "We have shared the design process with him, we have discussed process, material, shape, and sound," said Hossann. All the functional sounds such as the turning indicators and warning sounds.

As well as presenting the possibility of designing an EV with specific functional sounds, there could be a practical application. All electric cars sold in the European Union from 2019 will be required to make a sound at speeds up to 30 km/h (19 mph) to warn pedestrians of their presence.

The Peugeot team thinks this could be an opportunity to develop a specific sound signature for Peugeot electric cars. In addition, the Fractal design team has tapped into other value judgments that car buyers make about cars, such as the quality of the sound of the door when it closes. "That is why, for example on this car we have developed a sound for opening and closing the doors," said Hossann.

The designers also considered current personalization of cars, "Personalization today is a lot of adding things like stickers and paint. Tomorrow, you could have the same body, but just personalize the sound like a smartphone such as the door opening."

Sound has also been used to determine interior design features. "The best shape is an anechoic chamber. In fact we work with parametric design. We don't design just by drawing every line and making it look right," said Hossann. "We just work with an algorithm. We define a shape—it's a new way of conceiving things. This shape is impossible to produce by stamping, or with carbon fiber, or injection molding."

Hossann pointed to the inner door panel: "That's why we use a lot of 3D printing. More than 80% of the interior of the car is produced by 3D printing. Again this is possible because it is a concept car, but it's something that we believe could change the car industry in fact."

"At this stage it's difficult to have the speed and the volume", said Hossann, "But for example for small items like some décor, you can imagine in the future a

**Fractal is an electric urban coupe concept with Peugeot's latest generation i-Cockpit interior integrating sound.**

customer might be able to choose a specific area where they can add some specific décor and have this printed at the dealer and fitted to the car.

"We designed the car like a classical designer, but we almost designed the way of thinking about our car. We had a big discussion about the way of thinking about things and developing them. On a classical car you have a door panel, you add some foam to make the car more soundproof inside. Tomorrow we can imagine having an efficient shape for sound so that you don't need to add soundproofing material, so you can save weight." ■

*John Kendall*

# Peugeot seeks international appeal with Quartz concept

**Peugeot's Quartz concept is based on the PSA EMP2 platform powered by a high performance plug-in hybrid powertrain.**

Peugeot exhibited two concepts at the 2014 Paris Motor Show, the Exalt, a development of the Onyx concept seen at the Paris show in 2012 and Quartz, a crossover concept based on the European C-segment PSA EMP2 platform, powered by a plug-in hybrid drivetrain, producing a total of 500 hp (373 kW), developed by Peugeot Sport.

Quartz Design Manager Matthias Hossann says that the reason for combining a crossover with a high-performance powertrain was the need to design a concept with international appeal. Hossann spent five years working in PSA's design studio in China and was inspired to create a mix of performance car and SUV by his experience of the Chinese market.

Inside the car, Basalt stone is used extensively on the center console as well as on the doors and around the headlamps. Hossann explained, "We were looking for a more iconic material and we found the lava stone." The material is produced in volcanic eruptions and Peugeot claims that its properties are the same wherever it is found; so sourcing the material for target markets should not be difficult. Hossann says that the porous Basalt is both strong and light, so well suited to use on a car. "At Peugeot, we love to play with materials without processing them. The basalt is a good example. We just cut it but keep the material as it is. This is the kind of philosophy we would like to keep for the next generation."

Hossann said that the exterior could be separated into three parts, "The low part is clearly inspired by the SUV, then the body which is inspired by more sensual shapes, and then the cabin which we call the black diamond because we try to inject some "edge" on the glass. We've done this to make the silhouette easier to read, to control the light on the car.

"We made the cabin in a black transparent material because if you look at a classic car, you have the windshield, the roof, the rear window and the car looks like it is split into many parts. The idea with

**The glass roof flows into two separate spoilers at the rear of the Quartz.**

**The 23-in alloy wheels of the Quartz are constructed from thin material sections to reduce weight. Composite flaps direct cooling air to the brakes**

Quartz is to make something that looks really monolithic and really looks like a block." As the roof section flows into the rear wings the glass roof is formed into two separate spoilers designed to improve the aerodynamics.

The Quartz is fitted with 23-in alloy wheels, which as Hossann pointed out could introduce weight issues, "So the idea was to create something with a thin structure...for such a large wheel." These are fitted with composite flaps, which are designed to improve aerodynamic flow to help cool the brakes.

The body structure does not have a B-pillar and Peugeot has given the cars scissor doors, which are hinged high on the A-pillars and on the C-pillars for the rear. The vehicle uses a composite structure with bonded panels, and Peugeot claims reduced weight and high body stiffness, enabling the structure to function without a B-pillar.

For the interior, Peugeot claims to have produced the first digitally woven textile material to be used in a vehicle. The process is said to create large and complex components. The textile material is woven with polyester fiber produced from the plastic used to produce drinking water bottles. The material can be produced in thicknesses great enough to avoid the need for seat foam because the material is also compliant.

Power is provided by a 1.6-L turbocharged direct injection gasoline engine producing 270 hp (201 kW) and 330 N·m (243 lb·ft). The engine drives through a six-speed automatic transmission. An 85-kW electric motor also provides drive to the front wheels and draws on Peugeot's experience with its Hybrid4 system that helps to smooth gear changes. A second 85-kW motor drives the rear wheels. Both electric motors provide regenerative charging.

Three drive modes are available: ZEV, Road, and Race modes. In ZEV mode, the Quartz has a 50-km (31-mi) all-electric range. Both gasoline engine and front electric motor work together in Road mode, while in Race mode, both electric motors and the gasoline engine are used. Limited slip differentials are fitted front and rear.

Air suspension is used with McPherson type struts at the front and a multi-link arrangement at the rear. A camera-based system maintains ride height between 300 and 350 mm (11.8 and 13.8 in). The system is linked to a GPS navigation system to anticipate changes in gradient. ■

*John Kendall*

# Toyota reveals new platform, hybrid powertrain in C-HR concept

**The styling theme of the C-HR C-segment crossover hints at the future design direction for Toyota vehicles.**

The C-HR crossover concept is an important vehicle for Toyota, demonstrating the Japanese automaker's desire for more "engaging" design. Unique styling is a key attribute for a new vehicle, particularly in the highly competitive European C-segment market, said Karl Schlicht, Executive Vice President of Toyota Motor Europe, at the Paris Motor Show unveiling.

Significantly, the company's $ED^2$ (European Design Development) studio based in Southern France led the design for the concept, which introduces what Toyota calls a "diamond architecture" styling theme. The lower bodywork is "sculpted to represent the facetted surfaces of a highly durable, precision-cut gemstone," press materials expound.

Setting such florid descriptions aside, the styling hints at the future design direction for its vehicles, Toyota promises.

Key design elements in the front include a slim upper grille extending into the headlights that incorporate a unique 3-D treatment of the daytime running lights, as well as an aero-inspired floating front spoiler. From the side, the glasshouse tapers to the rear window, which along with the "spoiler-tipped roof line" creates the effect of a floating roof.

Beyond the design elements, the crossover concept is significant because it introduces a new platform with what Toyota claims will be a lower center of gravity and increased structural rigidity. Improved handling and a more "fun-to-drive" vehicle are

**The C-HR's glasshouse tapers to the rear window, which along with the "spoiler-tipped roof line" creates the effect of a floating roof.**

important bogeys for the C-HR, said a Toyota spokesman after the concept's Paris debut.

Smaller than the RAV4, the crossover concept is 4350 mm (171.3 in) long, 1850 mm (72.8 in) wide, and 1500 mm (59.0 in) tall, with a wheelbase of 2640 mm (103.9 in). (See chart for size comparison to the 2014 RAV4.)

The concept also employs a new full-hybrid powertrain that reportedly will deliver "significantly" improved fuel efficiency.

Questions to the spokesperson regarding the new hybrid system, platform specifics, and material usage were invariably met with "Unfortunately, I cannot provide any more details at this time."

He did offer, though, that the next-generation hybrid system is "certainly a level up from what we have in our current [hybrid] vehicles."

The European market is a major one for Toyota and Lexus hybrid sales. Through the first eight months of 2014, hybrid sales for the brands reached 107,000 units, representing 28% of the Toyota Group sales in Western Europe. ■

*Ryan Gehm*

# Cross Coupé GTE previews seven-seat SUV from VW Chattanooga

**The Volkswagen Cross Coupé GTE is 190.8 in (4847 mm) long, 68.3 in (1736 mm) tall, and 79.9 in (2030 mm) wide.**

Volkswagen officials have acknowledged that the company's sales have suffered in the U.S. from a lack of offerings in the rapidly growing crossover/SUV market, so it is scrambling to offer more models in that category.

The company just launched a refreshed Touareg built in Slovakia, but the more significant U.S.-sourced SUVs are coming soon. The efforts will be supported by a ramp-up of engineering staff to over 200 strong at a development and planning center in Chattanooga, TN, said Prof. Dr. Martin Winterkorn, CEO and Chairman of the Board of Management of Volkswagen AG at the 2015 North American International Auto Show (NAIAS) in Detroit.

At the end of 2016, the company will start production in Chattanooga, TN, of a seven-seat mid-size SUV. Then in 2017 it will begin producing an all-new long-wheelbase Tiguan, said Winterkorn.

In a preview of the new seven-seat model, Volkswagen unveiled the five-seat Cross Coupé GTE concept at the NAIAS. The acronym GTE applies to all VWs with plug-in hybrid drive, offering the characteristics of a grand tourer

**The plug-in hybrid Cross Coupé GTE concept is powered by a 3.6-L V6 gasoline engine and two electric motors.**

(GT, or Gran Turismo) with one or more electric motors and a battery (E, as in Electric). In Europe, the first GTE is already on the market.

The concept "is the ambassador of a new design language developed by Volkswagen for the U.S." said Klaus Bischoff, VW Chief Designer. "Numerous details hint at how we envisage a future production SUV model for North America."

It is the third concept vehicle previewing the production version of VW's upcoming mid-size SUV, following the CrossBlue from the 2013 NAIAS and the CrossBlue Coupé from 2013 Los Angeles, showing a range of SUV variants made possible by VW's modular transverse matrix (MQB) component set.

**The Cross Coupé GTE concept design team was led by Klaus Bischoff, Head of Volkswagen Design.**

The Cross Coupé GTE is 190.8 in (4847 mm) long, 68.3 in (1736 mm) tall, and 79.9 in (2030 mm) wide. The front features a high front end with dual LED headlights, indicators, and the radiator grille merged into a single design module. The daytime running lights will be a recognized signature in future production models, with four "LED light wings," two at the top in the radiator grille and two in the lower air inlet.

The MQB is flexible, allowing for conventional drive types (gasoline and diesel engines), plug-in hybrid, natural gas, and zero-emissions (pure electric drive and hydrogen fuel cell), so the production version of the Cross Coupé GTE could have drive options. However, VW chose from the mix a 3.6-L V6 gasoline engine producing 206 kW (276 hp) and 350 N·m (258 lb·ft) in conjunction with two electric motors delivering 40 kW and 220 N·m (162 lb·ft) in front and 85 kW and 270 N·m (199 lb·ft) in the rear. Total output of 265 kW (360 PS) provides a top speed of 130 mph (209 km/h) and acceleration to 60 mph (97 km/h) in 6.0 s. Fuel economy is 70 MPGe (combined) as measured by the U.S. EPA's standard adapted for hybrid vehicles with full use of the battery charge.

Electrical energy comes from a compact 14.1-kW·h lithium-ion battery pack housed in the concept's center tunnel.

The concept can be a zero-emissions electric vehicle for up to 20 mi (32 km) on a charge and can be driven in various operating modes: E-Mode, GTE, Hybrid, Off-road, and Battery Hold/Battery Charge.

In E-Mode, it can travel up to 20 mi (32 km) on pure electricity through the 85-kW rear axle electric motor. The V6 is decoupled from the drivetrain and shut down. Power electronics in the front engine compartment manage energy flow from and to the battery and electric motors. The 12-volt electrical system is supplied by a dc/dc converter.

**Dr. Heinz-Jakob Neußer, Member of the Board of Management of the Volkswagen brand with responsibility for 'Development', explains the technical details of the Cross Coupé GTE concept.**

The concept starts in Hybrid mode by default, operating as a full hybrid with regenerative braking and automatic use of the V6 and/or electric motor depending on the drive needs. Battery state-of-charge is kept constant unless the driver intervenes with a battery hold function.

When the driver takes his/her foot off the accelerator and the battery pack is sufficiently charged, the engine and all motors are shut down and disengaged from the drivetrain in Coasting mode. If the battery pack is not sufficiently charged, both electric motors work as generators and feed the energy generated during braking to the battery. In this case, too, the V6 is shut down and disengaged.

When just the gasoline engine is used, the concept is a purely front-wheel drive

GTE mode makes accelerator pedal, transmission, and steering characteristics sportier and the electric motors boost the V6 for full system output.

In Off-road mode, all four wheels are powered. If battery-pack state-of-charge is low, the front electric motor acts as a generator driven by the V6 to provide power for the rear axle motor. Since power to the rear axle flows by wire, not mechanically, VW refers to the arrangement as an "electric cardan shaft," and in Off-road mode, all-wheel drive is available even for a low state of charge battery.

The dual-mode Battery Hold (energy content of the battery stays constant) and Battery Charge (battery is charged while driving) mode is activated via a sub-menu in the infotainment system. It can force all-electric mode (E-Mode) in urban situations.

**Inside, the Cross Coupé GTE's instrument panel is fully electronic, with an Active Info Display in front of the driver and a touch screen center display.**

Inside, the Cross Coupé GTE's instrument panel is fully electronic, with a glossy black horizontal element having integrated Active Info Display in front of the driver and a touch screen center display.

The 12.3-in 1440 x 540 pixel Active Info Display digital instrument can also show navigation information and its graphics change also change depending on the drive mode. The 10.1-in infotainment touch screen features approach sensors for gesture control. The layout of the touchscreen graphics changes depend on driving mode.

The seat system has buckets in front and a rear bench that can be folded for a continuous cargo floor. Rear-seat entertainment is handled by attachments for tablets in the headrests of the front seats. ■

*Kevin Jost*

# VW's carbon-fiber XL Sport concept sports Ducati V2 motorcycle engine

**Like its XL1 sibling, the Volkswagen XL Sport is extremely aerodynamic. The concept has a Cd of 0.258 and its frontal area is 1.7 m² (18.3 ft²).**

Volkswagen refers to the XL Sport concept vehicle that it revealed at the 2014 Paris Motor Show as a "sister model" of the XL1 (see http://articles.sae.org/12588/ and http://articles.sae.org/11874/). As with most siblings, some similarities are evident—but the differences can be striking.

First, the similarities. Like the XL1, the XL Sport is extremely aerodynamic. The performance car concept has a drag coefficient of 0.258 (vs. 0.189 for the XL1) and its frontal area is 1.7 m² (18.3 ft²). Design features contributing to cooling and downforce include air curtains that direct air in the frontal area into specific channels, wheel arch ventilation, an optimized underbody, lift-reducing air ducts in the hood, an extendible rear spoiler similar to that on the Lamborghini Aventador, and adaptive waste heat vents incorporated in the rear hatch that can open automatically to cool the drive unit.

Another shared attribute is a lightweight monocoque (albeit differently shaped) made largely of carbon-fiber-reinforced polymer (CFRP), featuring slightly offset driver and passenger seats. Both cars sport wing doors that extend into the roof and are hinged at two points—low on the A-pillars and just above the windshield—allowing them to swing upward and slightly forward. Door windows are made of polycarbonate, and the windshield is a special type of thin glass.

At 890 kg (1962 lb), the XL Sport weighs more than the XL1 [795 kg (1753 lb)]. The concept is longer and wider than the XL1, and also has an increased wheelbase (see chart for dimensions). Whereas the XL1 tapers toward the rear, the XL Sport maintains its width front to back. The sport concept inherits the XL1's e-mirrors—cameras integrated in the doors that send images of the surroundings to two interior displays.

Now to the fun part: the differences. Most notable is the powertrain. Under the rear hatch of the XL Sport is what VW claims is the world's most powerful two-cylinder engine: the Ducati 1199 Superleggera's V2 engine. The 1199-cm³ DOHC powerplant generates 147 kW (197 hp) and 134 N·m (99 lb·ft), can propel the concept car to 270 km/h (168 mph), and revs to 11,000 rpm. The unit was slightly modified for use in the XL Sport, according to VW, but is basically the same as the motorcycle engine.

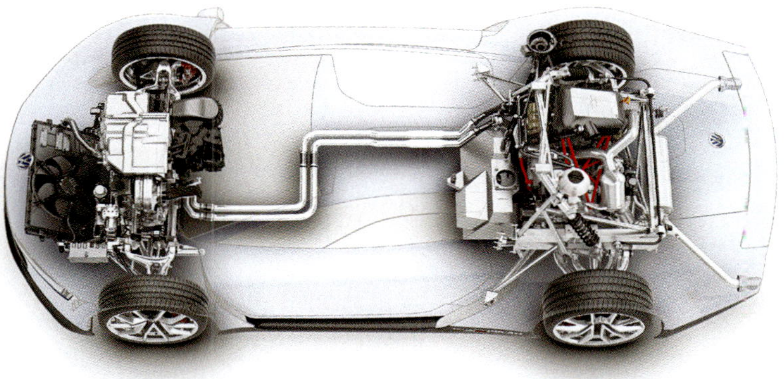

True to its moniker, the XL Sport gets a significantly redesigned, racing-inspired chassis incorporated in a high-strength steel spaceframe.

Wing doors extend into the roof and are hinged at two points—low on the A-pillars and just above the windshield—allowing them to swing upward and slightly forward.

The rear hatch conceals the Ducati V2 engine, the 7-speed DSG, and 107 L (3.78 $ft^3$) of luggage space.

The Ducati engine's high speeds are possible thanks to titanium connecting rods, its bore/stroke ratio of 112 mm (4.41 in)/60.8 mm (2.39 in), respectively, and the "exceptionally short" crankshaft stroke associated with it. The two four-valve cylinders are arranged 90° to each other and feature a desmodromic valve control system (positive valve closure) typical of high-revving Ducati engines. Other features include a magnesium-alloy clutch, cylinder head and oil pan covers, the two disks of the throttle valves, and the two injectors per cylinder.

The XL Sport's newly developed intermediate gearbox reduces engine speeds by a factor of 1.86. Torque is transmitted to the rear axle via a seven-speed dual-clutch gearbox (DSG).

The car accelerates from 0-100 km/h (0-62 mph) in 5.7 s.

True to its moniker, the XL Sport gets a significantly redesigned, racing-inspired chassis incorporated in a high-strength steel spaceframe. Its double wishbone front axle sees the dampers connected below in a pull-rod configuration, and a double wishbone rear axle has the dampers connected above in a push-rod setup.

High-speed 205/40 R18 (front) and 265/35 R18 (rear) tires wrap forged magnesium wheels, which are reportedly 23.9 kg (52.7 lb) lighter than aluminum wheels combined. The brake system features ceramic discs. ■

*Ryan Gehm*

SAE Eye on Engineering Quick Take:

## ZF's Smart Urban Vehicle concept

Watch the video at https://youtu.be/-uOTO8H51zY.

City dwellers need vehicles that are tailored to the tight confines of urban driving. In this episode of SAE Eye on Engineering, Senior Editor Lindsay Brooke looks at ZF's Smart Urban Vehicle. SAE Eye on Engineering can be viewed at https://youtu.be/-uOTO8H51zY. It also airs in audio-only form Monday mornings on WJR 760 AM Detroit's Paul W. Smith Show.

Access archived episodes of SAE Eye on Engineering at www.sae.org/magazines/podcasts.